Geopolymers as Sustainable Surface Concrete Repair Materials

The progressive deterioration of concrete surface structures is a major concern in construction engineering that requires precise repairing. While a number of repair materials have been developed, geopolymer mortars have been identified as potentially superior and environmentally friendly high-performance construction materials, as they are synthesized by selectively combining waste materials containing alumina and silica compounds which are further activated by a strong alkaline solution. *Geopolymers as Sustainable Surface Concrete Repair Materials* offers readers insights into the synthesis, properties, benefits and applications of geopolymer-based materials for concrete repair.

- Discusses manufacturing and design methods of geopolymer-based materials
- Assesses mechanical strength and durability of geopolymer-based materials under different aggressive environmental conditions
- Characterizes the microstructure of these materials using XRD, SEM, EDX, TGA, DTG and FTIR measurements
- Describes application of geopolymer-based materials as surface repair materials
- Compares environmental and cost benefits against those of traditional OPC and commercial repair materials

This book is written for researchers and professional engineers working with concrete materials, including civil and materials engineers.

Emerging Materials and Technologies

Series Editor
Boris I. Kharissov

Heterogeneous Catalysis in Organic Transformations
Varun Rawat, Anirban Das, Chandra Mohan Srivastava

2D Monoelemental Materials (Xenes) and Related Technologies: Beyond Graphene
Zongyu Huang, Xiang Qi, Jianxin Zhong

Atomic Force Microscopy for Energy Research
Cai Shen

Self-Healing Cementitious Materials: Technologies, Evaluation Methods, and Applications
Ghasan Fahim Huseien, Iman Faridmehr, Mohammad Hajmohammadian Baghban

Thin Film Coatings: Properties, Deposition, and Applications
Fredrick Madaraka Mwema, Tien-Chien Jen, and Lin Zhu

Biosensors: Fundamentals, Emerging Technologies, and Applications
Sibel A. Ozkan, Bengi Uslu, and Mustafa Kemal Sezgintürk

Error-Tolerant Biochemical Sample Preparation with Microfluidic Lab-on-Chip
Sudip Poddar and Bhargab B. Bhattacharya

Geopolymers as Sustainable Surface Concrete Repair Materials
Ghasan Fahim Huseien, Abdul Rahman Mohd Sam, and Mahmood Md. Tahir

Nanomaterials in Manufacturing Processes
Dhiraj Sud, Anil Kumar Singla, Munish Kumar Gupta

For more information about this series, please visit: https://www.routledge.com/Emerging-Materials-and-Technologies/book-series/CRCEMT

Geopolymers as Sustainable Surface Concrete Repair Materials

Ghasan Fahim Huseien, Abdul Rahman Mohd Sam, and Mahmood Md. Tahir

CRC Press
Taylor & Francis Group
Boca Raton London New York

CRC Press is an imprint of the
Taylor & Francis Group, an **informa** business

First edition published 2023
by CRC Press
6000 Broken Sound Parkway NW, Suite 300, Boca Raton, FL 33487-2742

and by CRC Press
4 Park Square, Milton Park, Abingdon, Oxon, OX14 4RN

CRC Press is an imprint of Taylor & Francis Group, LLC

© 2023 Taylor & Francis Group, LLC

Reasonable efforts have been made to publish reliable data and information, but the author and publisher cannot assume responsibility for the validity of all materials or the consequences of their use. The authors and publishers have attempted to trace the copyright holders of all materials reproduced in this publication and apologize to copyright holders if permission to publish in this form has not been obtained. If any copyright material has not been acknowledged please write and let us know so we may rectify in any future reprint.

Except as permitted under U.S. Copyright Law, no part of this book may be reprinted, reproduced, transmitted or utilized in any form by any electronic, mechanical or other means, now known or hereafter invented, including photocopying, microfilming and recording, or in any information storage or retrieval system, without written permission from the publishers.

For permission to photocopy or use material electronically from this work, access www.copyright. com or contact the Copyright Clearance Center, Inc. (CCC), 222 Rosewood Drive, Danvers, MA 01923, 978-750-8400. For works that are not available on CCC please contact mpkbookspermissions@tandf.co.uk

Trademark notice: Product or corporate names may be trademarks or registered trademarks and are used only for identification and explanation without intent to infringe.

ISBN: 978-1-032-00299-6 (hbk)
ISBN: 978-1-032-00305-4 (pbk)
ISBN: 978-1-003-17361-8 (ebk)

DOI: 10.1201/9781003173618

Typeset in Times
by codeMantra

Contents

Preface ix
About the Authors xi

Chapter 1 Concrete Durability and Surface Deterioration 1

    1.1   Introduction 1
    1.2   Concrete Durability 2
    1.3   Concrete Surface Deteriorations 3
    1.4   Causes of Concrete Degradation 3
    1.5   Existing Surface Repair Materials 4
    1.6   Waste Materials–Based High-Performance Geopolymers 6
    1.7   Summary 7
    References 7

Chapter 2 Geopolymer as Emerging Repair Materials 11

    2.1   Introduction 11
    2.2   Geopolymer Mortar 13
    2.3   Geopolymer Mortar as Repair Materials 15
    2.4   Geopolymer Performance 16
    2.5   Durability and Sustainability 20
    2.6   Economic Feasibility 21
    2.7   Environment Suitability and Safety Features 22
    2.8   Merits and Demerits of Geopolymer as Repair Material 23
    2.9   Summary 23
    References 24

Chapter 3 Manufacturing Geopolymer: Materials and Mix Design 31

    3.1   Introduction 31
    3.2   Fly Ash-Based Geopolymer Binder 32
            3.2.1   Effect of FA on Workability and Strength Properties 32
            3.2.2   Effect of FA on Durability of Geopolymer 33
    3.3   Palm Oil Fuel Ash 33
            3.3.1   Effect of POFA on Workability and Strength Properties 34
            3.3.2   Effect of POFA on Durability of Geopolymer 35
    3.4   Ground Blast Furnace Slag 35
            3.4.1   Effect of GBFS on Workability and Strength Properties 35
            3.4.2   Effect of GBFS on Durability of Geopolymer 36

	3.5	Ceramic Wastes	37
		3.5.1 Effect of Ceramic Wastes on Workability and Strength Properties	37
		3.5.2 Effect of Ceramic Waste on Durability of Geopolymer	38
	3.6	Alkaline Activator Solutions	38
		3.6.1 Workability and Strength Performance	39
		3.6.2 Effect of Solution on Durability of Geopolymer	41
	3.7	Characteristics of Various Geopolymers	42
	3.8	Geopolymer Mix Design	42
	3.9	Summary	44
	References		44
Chapter 4	Factors Effect on the Manufacturing of Geopolymer		51
	4.1	Introduction	51
	4.2	Fresh Properties of Geopolymer	52
	4.3	Compressive Strength	54
		4.3.1 Effect of Calcium Content	54
		4.3.2 Effect of Alkaline Solution Characterization	56
		4.3.3 Effect of Aggregate-to-Binder Ratio	58
		4.3.4 Effect $H_2O:Na_2O$ Ratio	58
		4.3.5 $SiO_2:Na_2O$ Ratio Effect	59
	4.4	Bond Strength	59
		4.4.1 Effect Calcium Content	59
		4.4.2 Effect of Alkaline Activator Solution	61
		4.4.3 Effect of Silicate-to-Aluminium Ratio	62
		4.4.4 Effect of Solid-to-Liquid Ratio	62
		4.4.5 Effect of Curing Humidity	63
		4.4.6 Effect of $SiO_2:K_2O$ Ratio	63
		4.4.7 Bond Strength at Elevated Temperatures	64
	4.5	Flexural Strength	64
	4.6	Drying Shrinkage	65
	4.7	Abrasion–Erosion Resistance	67
	4.8	Microstructures	68
	4.9	Failure Mode and Interface Zone	71
	4.10	Summary	72
	References		73
Chapter 5	Performance Criteria of Geopolymer as Repair Materials		77
	5.1	Introduction	77
	5.2	Geopolymer Binder	78
	5.3	Geopolymer Mix Design	81
	5.4	Workability Performance	83

	5.5	Compressive Strength Performance 87
	5.6	Splitting Tensile Strength 89
	5.7	Flexural Strength 90
	5.8	Bond Strength Performance 91
	5.9	Summary 92
	References 93
Chapter 6	Compatibility of Geopolymer for Concrete Surface Repair 97
	6.1	Introduction 97
	6.2	Geopolymer Preparation 99
	6.3	Workability of Fresh GPMs 102
	6.4	Strength Performance 104
	6.5	Slant Shear Bonding Strength 110
	6.6	Thermal Expansion Coefficient 111
	6.7	Third-Point Loading Flexural 113
	6.8	Bending Stress 116
	6.9	Summary 117
	References 118
Chapter 7	Effects of Aggressive Environments on Geopolymer Performance as Repair Materials 123
	7.1	Introduction 123
	7.2	Geopolymer Ternary Blended 125
	7.3	Procedures of Geopolymer Tests 129
	7.4	Compressive Strength Performance 130
	7.5	Bond Strength of Geopolymer 132
	7.6	Effect of Sulphuric Acid Attack 134
	7.7	Geopolymer Resistance to Sulphate Attacks 138
	7.8	Effect of Elevated Temperatures 141
	7.9	Summary 143
	References 145
Chapter 8	Performance Evaluation of Geopolymer as Repair Materials Under Freeze–Thaw Cycles 149
	8.1	Introduction 149
	8.2	Mix Design 151
	8.3	Porosity 152
	8.4	Surface Abrasion Resistance 154
	8.5	Freezing–Thawing Cycle Resistance 155
	8.6	Drying Shrinkage 160
	8.7	Wet–Dry Cycle Resistance 161
	8.8	Summary 163
	References 164

Chapter 9 Methods of Evaluating the Geopolymer Efficiency as Alternative Concrete Surface Repair Materials Compared to Commercials Products 167

    9.1   Introduction 167
    9.2   Causes of Concrete Surface Degradation............................. 169
    9.3   Commercial Repair Materials 170
            9.3.1   Cement-Based Materials 171
            9.3.2   Polymer-Modified Cement-Based Materials 172
            9.3.3   Epoxy-Based Materials 172
    9.4   Selection of Repair Materials 173
    9.5   Development of Geopolymer as Repair Materials 173
    9.6   Efficiency Evaluation of Geopolymer as Repair Materials... 175
    9.7   Summary 176
    References 178

Chapter 10 Sustainability of Geopolymer as Repair Materials 179

    10.1   Introduction 179
    10.2   Geopolymer Preparation 181
    10.3   Strength Performance 183
    10.4   Life Cycle Assessment.. 187
    10.5   Modified LCA with Respect to CS and Durability 193
    10.6   ANN for Estimating CO_2 Emission and EE........................ 195
            10.6.1   Rationale... 195
            10.6.2   Cuckoo Optimization Algorithm 195
            10.6.3   Generation of Training and Testing Data Sets 196
            10.6.4   Model Predictions and Results.............................. 199
    10.7   Summary 202
    References 203

Index 207

Preface

Geopolymers as Sustainable Surface Concrete Repair Materials is the first edition of the book comprising 10 chapters under the series Emerging Materials and Technologies. Each of these chapters discusses one of the applications of geopolymer binder as a surface concrete repair material in the concrete industry. Utilization of geopolymers as repair materials in the concrete industry not only improves the engineering properties but it was also found to help reduce landfill and environment problems and achieve the sustainability goals.

The first chapter deals with the general introduction on concrete durability, deterioration cases and performance in aggressive environments. Industrial and agriculture waste–based geopolymer binder and the effect of chemical, physical and mineral raw materials on geopolymerization process are widely discussed in Chapter 2. In Chapter 3, we review the factors' effect on geopolymer mix design such as alkaline activator solution molarity and modulus, binder and filler contents, water-to-binder ratio and curing regime. The performance criteria, compatibility between geopolymer and concrete substrate, resistance to sulphuric acid and sulphate attacks, and wet–dry and thaw–freeze cycles are presented in Chapters 4–8. Topics regarding environmental and sustainable benefits and future utilization are presented in Chapters 9 and 10.

The utilization of waste-based geopolymer binders in sustainable surface concrete repair materials production and construction industry whose growth knows no boundaries has been studied. Mounting evidence of worldwide interest suffices the need to produce a collective anthology of a wide variety of sustainable repair materials for current and future applications.

Ghasan Fahim Huseien
Abdul Rahman Mohd Sam
Mahmood Md. Tahir

About the Authors

Dr Ghasan Fahim Huseien is a researcher at the Department of the Built Environment, College of Design and Engineering, National University of Singapore, Singapore. He received his Ph.D. degree from the University Technology Malaysia in 2017. He is a member of Concrete Society of Malaysia and American Concrete Institute. He has over 7 years of experience in Applied Research and Development and has up to 10 years of experience in manufacturing smart materials for sustainable building and smart cities. He has expertise in Advanced Sustainable Construction Materials covering Civil Engineering, Environmental Sciences and Engineering and Chemistry. He authored and co-authored more than 80 publications and technical reports, 5 books and 21 book chapters, and participated in more than 30 national and international conferences. His past experience in projects includes the application of nanotechnology in construction and building materials, self-healing technology, and geopolymer as sustainable and eco-friendly repair materials in the construction industry.

Dr Abdul Rahman bin Mohd. Sam is an Associate Professor at the School of Civil Engineering, Faculty of Engineering, Universiti Teknologi Malaysia (UTM) where he has been a faculty member since 1988. Currently, he is the director of UTM Construction Research Center (UTM CRC), Institute for Smart Infrastructure and Innovative Construction, UTM. He obtained his B.Sc. (Hons) Civil Engineering from California State University, Sacramento, United State of America; M.Sc. from Universiti Teknologi Malaysia, Malaysia; and Ph.D. from University of Sheffield, United Kingdom. His research interests include concrete durability, smart concrete materials, fibres in concrete, self-healing concrete, self-curing concrete, sustainable construction materials and environmental engineering. He authored and co-authored more than 180 publications and technical reports. Some of his research output was published in Q1 & Q2 with high impact factor journals including *Journal of Cleaner Production*, *Construction and Building Materials*, *Journal of Building Engineering, Materials, Sustainability* and other index journals. Up to now, Dr Abdul Rahman and his research team members have won more than 25 gold, silver and bronze medals in various local and international competitions related to research products. His H-index is 24 and i-index is 44.

Prof Ir. Dr Mahmood Md. Tahir joined UTM in May 1985 as an Assistant Lecturer "A" after completing his first degree from the University of Iowa, USA. He taught diploma students from May 1985 to 1988 before pursuing his master's degree at the University of Nebraska–Lincoln, USA. After completing his master's degree (M.Sc.) in Structure Engineering, he continued to serve FKA, UTM from 1989 to 1993. He continued his study for Ph.D. in January 1994 at the University of Warwick, U.K. and managed to complete his study in May 1997. During his involvement in research and publication for the last 20 years, he has published more than 100 indexed journals and more than 80 conference papers. His main areas of research are steel structures, composite structures and concrete structures. He also has registered as a member of Institute Engineer Malaysia (MIEM). At present, he is a Senior Director at the Institute for Smart Infrastructure and Innovative Construction (ISIIC).

1 Concrete Durability and Surface Deterioration

1.1 INTRODUCTION

Over the years, Ordinary Portland Cement (OPC) has been widely employed as concrete binder and various building substances worldwide. It is known that large-scale manufacturing of OPC causes serious pollution in the environment in terms of considerable amount of greenhouse gases emission [1,2]. The OPC production alone is accountable for nearly 6%–7% of total CO_2 emissions as estimated by International Energy Agency [3]. In fact, among all the greenhouse gases, approximately 65% of the global warming is ascribed to the CO_2 emission. It was predicted that the mean temperature of globe could raise by approximately 1.4°C–5.8°C over the next 100 years [4]. Globally, in the present backdrop of CO_2 emissions–mediated climate change, the sea level is expected to rise, and the subsequent occurrence of natural disasters will cause huge economic loss [5]. On top of this, the greenhouse gases, such as CO_2, SO_3 and NO_X, emitted from the cement-manufacturing industries can cause acid rain and damage the soil fertility [6]. Generally, the industrial consumption of raw materials is around 1.5 tonnes per each tonne of OPC production [7]. To surmount such problems, scientists, engineers and industrial personnel have been continuously dedicating many efforts to develop novel construction materials to achieve alternate binders [8].

Several studies [9–11] reported that the durability of concrete becomes low in aggressive environments, which leads to earlier deterioration during its life service. Wang et al. [12] reported that the cost of restorations and rehabilitations was close to or even exceeded the cost of a new construction. The surfaces of concrete structures, such as sidewalks, pavements, parking decks, bridges, runways, canals, dykes, dams, and spillways, deteriorate progressively due to a variety of physical, chemical, thermal and biological processes. Actually, the performance of concrete compositions can greatly be affected by the improper usage of substances, and physical and chemical conditions of the environment [10,13,14]. The immediate consequence is the anticipated need of maintenance and execution of repairs [15,16]. To overcome these issues, the researchers made dedicated efforts to develop different types of repair materials, with or without OPC, such as the emulsified epoxy mortars (EMs), sand epoxy mortars and polymer-modified cement-based mortars (PMCMs). The main aim is to develop efficient and durable materials for repairing damaged concrete structures. However, the variation between the results obtained by different researchers can be attributed to the difference in the raw materials, specimen geometry and test methods. For the construction purpose, the geopolymer (GP) pastes and mortars are the newly introduced binders with much higher resistance against severe climatic conditions [17,18]. In the past, intensive efforts have been made to develop binders with high performance as the sustainable construction materials [19–22].

DOI: 10.1201/9781003173618-1

1.2 CONCRETE DURABILITY

Characteristically, the serviceability of construction materials has considerable economic significance, particularly with modern infrastructures and components. For urbanization, concrete materials that are greatly exploited must meet the standard codes of practical requisites related to strength and durability [23]. For instance, poor plan, low capacity or overload, faulty material design and structures, wrong construction practices or insufficient maintenance and lack of engineering knowledge can often diminish the service lifespan of concrete under operation [24]. In the construction industries, fast declination of concrete structures being a major setback necessitates additional improvement. A variety of physical, chemical, thermal and biological processes are responsible for the progressive deterioration of concrete structures during their service [10,25]. Concrete performance is greatly affected by improper usage, and physical and chemical conditions of the surrounding environment (Figure 1.1). It is verified that both external and internal factors, involving physical, chemical or mechanical actions, are often responsible for the deterioration of concrete structures.

Mechanical damage of a concrete structure occurs due to different reasons, such as impact, abrasion, cracking, erosion, cavitation or contraction. Chemical actions that cause the declination of concretes are carbonation, reactions associated with alkali and silica or alkali and carbonate, and efflorescence. Moreover, outside attacks of chemicals happen primarily due to CO_2, Cl_2, SO_4 as well as several other liquids and gases generated by the industries. Physical causes of deterioration include the effects of high temperature or differences in the thermal expansion of aggregate and of the hardened cement paste. Another reason of deterioration is the occurrence of alternating freezing and thawing of concrete and the associated action of the de-icing salts. Physical and chemical processes of deterioration often act in a synergistic way including the influence of sea water on concrete. Poor durability performance of OPC in an

FIGURE 1.1 Degradation of concrete structures due to the influence of aggressive chemicals.

aggressive acidic or sulphate (especially in marine) environment is caused due to the existence of calcium complexes. These calcium complexes readily dissolve in acidic atmosphere, leading to enhanced porosity and thus fast deterioration [26].

In many parts of the world, OPC structures that are existing for many decades are facing rapid deterioration [27]. Definitely, the permanence of OPC is linked with the nature of concrete constituents, where CaO of 60%–65% and the hydration product $Ca(OH)_2$ of nearly 25% are responsible for fast structural decay. Several studies indicated that the fast reaction of $Ca(OH)_2$ with acidic surroundings causes OPC to be deprived of water, leading to acid fusion and weakening of resistance against aggressive attacks. On top of these, the intense reaction of evolved CO_2 with $Ca(OH)_2$ contributes to rapid corrosion of the concretes containing OPC [27]. The safety, service life, permanence and lifespan of the mix design of concretes are considerably influenced by the crack development and subsequent erosion. These distinguished drawbacks of OPC-based concretes drove researchers to enhance the properties of conventional OPC by adding pozzolanic materials, polymer and nanomaterials so that it becomes more sustainable and endurable. The immediate consequence for affected concrete structures is the anticipated need of maintenance and execution of repairing [28]. Thus, there is a renewed interest for the development of sustainable concrete to solve all these existing shortcomings taking harsh environmental conditioning and durability into account.

1.3 CONCRETE SURFACE DETERIORATIONS

For the repair and maintenance, several expensive surface repair mortars are easily available commercially. They are constantly being used without prior laboratory testing. Earlier, many mortars are based on cement; cementitious mortars (CMs) modified with polymers such as acrylics and styrene butadiene rubbers (SBRs), sand epoxy and emulsified epoxy have been developed to repair the damaged concrete surfaces. These repair materials are often sold in the market with the promise of achieving wonderful results [29].

Information on most of these commercially available products has always been inadequate, and thus the manufacturers do not have detailed information on the resistance of such mortars under adverse atmospheric conditions that exist in many parts of the globe. Even though some data on the performance of these repair materials are provided by the suppliers and manufacturers, the values are generally given based on the laboratory ambient temperature of $21°C \pm 1°C$. Furthermore, the practising engineers find difficulties to select the right product for a particular repairing purpose. Certainly, there is a need to select appropriate materials for repairing deteriorated surfaces of various concrete structures.

1.4 CAUSES OF CONCRETE DEGRADATION

It is verified that both external and internal factors, involving physical, chemical or mechanical actions, are responsible for the deterioration of the concrete. Mechanical damage occurs due to different reasons such as impact, abrasion, cracking, erosion, cavitation or contraction. Chemical actions that cause the declination of concretes are carbonation, reaction associated with alkali and silica, alkali and carbonate as well as efflorescence. Moreover, outside attacks of chemicals happen primarily due to CO_2, Cl_2, SO_4 as well as several other liquids and gases generated by the industries. Physical causes of deterioration include the effects of high temperature or differences in the thermal expansion of aggregate and of the hardened cement paste. Other reason of deterioration is the occurrence of alternating freezing and thawing of concrete and the associated action of the de-icing salts. Physical and chemical processes of deterioration often act in a synergistic way including the influence of sea water on concrete.

The attack on the external surface and internal cracks are the two main reasons of concrete degradation. In the former, processes such as corrosion, cavity formation, abrasion and scaling are involved. The latter mainly involves the effect of rapid humidity or temperature changes, crystallization pressure, structure loading and exposure to extreme temperatures (freezing, fire). Climatic and environmental factors are the main causes of concrete degradation. The deformation of concretes happens due to climate changes and exposure to various types of destructive chemicals. Low-temperature and high-humidity conditions can easily induce internal cracks due to freezing and thawing, de-icing salt spill out, reinforced erosion, and alkaline-aggregated chemical reactions. Conversely, at humid and hot climates, the severe attack of water and alkaline-aggregated chemical reactions are common. Besides, at dry climatic conditions, the carbonation event occurs. In the marine environment, the most common cause of degradation is attack by sea water (containing sulphates). Moreover, degradation may also happen due to erosion, glacial abrasion and freezing–thawing. On top of these, other factors causing concrete degradations may include structural design, placement, maintenance and characteristics of concrete materials. Thus, synthesis of superior concrete materials with new compositions and their subsequent characterization are demanded to protect them from external and internal degradations.

1.5 EXISTING SURFACE REPAIR MATERIALS

For the repair and maintenance of degrading concrete structures, many surface repair mortars are commercially available. However, they are constantly used before being tested in the laboratory. In the last few years, several concrete repair materials and methods have been developed. Sales representatives selling repair materials all promise wondrous results with their products. Information on these products has always been scarce, and manufacturers have been unable to supply specific data on these mortars' resistance capacities to the harsh conditions found in many parts of the globe. Even if data are available, it is normally for room temperature conditions and is therefore of very little value for structures exposed to severe hot and cold climatic conditions. Some experts also estimate that up to half of all concrete repairs fail. Interestingly, many of these *wonderful* materials do not perform the repair as desired, and thus concrete repairs remain tricky. There are few engineers who have adequate knowledge of concrete repair, and contractors with experience in concrete repair are scarce too.

TABLE 1.1
Types of Existing Repair Materials

Cement-Based Materials	Polymer-Modified Cement-Based Materials	Resinous Materials
Portland cement	Styrene butadiene rubber acrylic mortar (SBRAM)	Epoxy mortar (EM)
Magnesium phosphate mortar (PCMPM)	Vinyl acetate–modified mortar (VAMM)	Polyester mortar (PM)
High-alumina cement (HAC)		

Some laboratory tests are conducted prior to the repair job. Tests such as bond strength, abrasion–erosion resistance, shrinkage–expansion, compressive strength and thermal compatibility with base concrete are usually carried out to evaluate and compare the performance of various types of repair materials [30]. Besides, other tests that are performed include the freezing–thawing cycles and permeability. Although all these tests are considered important and essential, different views and opinions exist in the literature regarding their degree of importance. Commercially available surface repair materials are broadly classified into three primary groups: CMs, PMCMs and resinous mortars [31,32]. Table 1.1 depicts further subdivision of these groups.

Several degraded concrete structures all around the globe are progressively encountering problems involving immediate repairing or treatment [33–36]. Al-zahrani et al. [37] have reported the restoration of concrete structures in terms of their durability and life cycles during service. It was acknowledged that this repair should be finished rapidly for civic expediency. Presently, many concrete repairing systems are accessible commercially which are mostly based on OPC, polymers and latex, and all of them are broadly employed (Table 1.1). Nevertheless, when cementitious systems are implemented in concrete structures for repairing, cracks are often developed due to temperature variations generated from high heat of hydration at early age [38]. Additionally, they suffer from other limitations associated with the lack of early age strength achievement with appropriate workability [39].

Several studies [31,40–43] revealed that the existing surface concrete repair materials display good mechanical performance, such as compressive strength and bond, with substrate concrete and low performance under aggressive environmental conditions. Mirza et al. [10] have studied the durability of more than 45 various types of commercial repair materials, such as OPC, SBR, polymer materials and EMs. They reported that in severe climatic conditions, most commercial repair materials display low performance with regard to abrasion–erosion resistance, shrinkage–expansion, wetting–drying cycles, permeability and freezing–thawing cycles. Cusson and Mailvaganam [32] have reported that the epoxy and polymer repair materials display high compressive and bond strength and very low resistance to elevated temperatures (less than 300°C). CMs exhibit low performance compared to EMs and polyester mortars especially for early compressive and bond strength. Rare commercial repair materials show good durability performance but are very costly [10]. Morgan et al. [44] acknowledged that the compatibility between

TABLE 1.2
Structural Compatibility of Repair Mortars in Terms of General Requisites [44,45]

No.	Properties	Relation Between the Repair Mortar (Rp) and the Concrete Substrate (NC)
1	Strength in compression, tension and flexure	Rp ≥ NC
2	Modulus in compression, tension and flexure	Rp = NC
3	Poisson's ratio	Dependent on modulus and type of repair
4	Coefficient of thermal expansion	Rp = NC
5	Adhesion in tension and shear	Rp ≥ NC
6	Curing and long-term shrinkage	Rp ≥ NC
7	Strain capacity	Rp ≥ NC
8	Creep	Dependent on whether creep causes desirable or undesirable effects
9	Fatigue performance	Rp ≥ NC

the concrete substrate and repair mortars must meet certain requirements such as compressive strength, tensile, flexural and adhesion strength. These requirements in repair mortars are listed in Table 1.2.

1.6 WASTE MATERIALS–BASED HIGH-PERFORMANCE GEOPOLYMERS

Most of the commercial repair materials owing to their low durability and sustainability perform poorly under severe hot and cold climatic conditions. Although few epoxy repair materials display good performance, they are somewhat costly. Geopolymer mortars (GPMs) prepared from the waste materials with high content of aluminium silicate and alkaline activator solution have emerged as a leading repair material. Geopolymeric binders are preferred because they generate 70%–80% lesser CO_2 with remarkably reduced greenhouse gas emissions than Portland cement. However, new binders are prerequisite for enhanced durability performance, better sustainability, reduced cost and environmental affability.

Currently, intensive researches on the GP as emerging construction material have been undertaken, where most studies revealed that an elevated concentration of sodium hydroxide and high ratio of sodium silicate (Na_2SiO_3) to NaOH ≥ 2.5 are preferred for the production of high-performance GPM. Sodium silicate is known to impact negatively on the environment. Besides additional cost, high concentration of sodium hydroxide has negative effects on the environment and remains hazardous to the workers. High molarity of sodium hydroxide and enriched sodium silicate in alkaline solution content are the major problems for the usage of GPM as new construction materials, especially for repairing. This is a serious concern for the environmental safety because it is a mineral-based material with relatively high demand for sodium silicate during synthesis. These deficiencies caused by alkaline solution limit the diversified use of GP in the construction industry.

Several studies are carried out on the materials containing calcium compounds, especially ground-granulated blast furnace slag (GBFS). However, most of the investigations used high-volume contents of corrosive Na_2SiO_3 and/or NaOH to achieve GP products, which posed health and safety issues of workers during handling. Davidovits et al. [46] proposed a simple approach to produce cheap GP with improved mechanical properties by overcoming thermally activated processes and promoting an easy management. So far, no study evaluated the mechanical properties of such comprehensible GPMs. Most of the researchers have analysed the mineralogy and microstructure properties of GPMs.

Palomo et al. [47] developed two models to understand the binding characteristics of GPMs with alkaline solution activation. The first model concerns with the mild alkaline solution activation of silicon (Si) plus Ca substances including GBFS to produce C-A-S-H gels as the main product. The second model deals with the alkaline solution activation of Si plus Al substances including FA and MK that need a robust alkali solution to produce N-A-S-H gels as the major outcome. Therefore, a potential production procedure of GPMs needs to be developed where low alkaline solution concentration (low sodium hydroxide molarity and low amount of sodium silicate) must be used by combining the effect of slag, high aluminium and silicate content materials, including fly ash (FA), palm oil fuel ash (POFA) and wastes ceramic materials (WCM), with varying ratios of $SiO_2:Al_2O_3$, $CaO:SiO_2$ and $CaO:Al_2O_3$. Consequently, the present study intends to develop an environmentally friendly and low-cost GPM with a broad range of applications in the construction industry. Figure 1.1 illustrates the existing research gaps in the form of problem statement that needs to be bridged together with the possible solution strategies.

1.7 SUMMARY

As mentioned earlier, this book intends to generate new information on the use of multi-blend GPs by means of systematic methods of sample preparation from waste materials economically, appropriate and careful materials characterizations and subsequent data analyses valuable for the progress of standard specifications of multi-blend GPMs towards diversified realistic applications. This generated knowledge is expected to provide to the advancement of environmentally amiable and inexpensive GPMs for a broad array of usages in the building sectors. This would be greatly advantageous for sustainable development of construction repair materials, where waste disposal problems towards the land filling can be avoided and minimized. The new findings of this book are believed to render a basis for further studies and better knowledge on the behaviour of a multi-blend GPMs obtainable from the waste materials in a cheap and environmentally affable manner.

REFERENCES

1. Duxson, P., et al., The role of inorganic polymer technology in the development of 'green concrete'. *Cement and Concrete Research*, 2007. **37**(12): pp. 1590–1597.
2. Rashad, A., et al., Hydration and properties of sodium sulfate activated slag. *Cement and Concrete Composites*, 2013. **37**: pp. 20–29.

3. Palomo, Á., et al., Railway sleepers made of alkali activated fly ash concrete. *Revista Ingeniería de Construcción*, 2011. **22**(2): pp. 75–80.
4. Rehan, R. and M. Nehdi, Carbon dioxide emissions and climate change: policy implications for the cement industry. *Environmental Science & Policy*, 2005. **8**(2): pp. 105–114.
5. Stern, N.H., *The economics of climate change: the Stern review*. 2007: Cambridge University press.
6. Zhang, Y., et al., Aspen Plus-based simulation of a cement calciner and optimization analysis of air pollutants emission. *Clean Technologies and Environmental Policy*, 2011. **13**(3): pp. 459–468.
7. Rashad, A.M., Properties of alkali-activated fly ash concrete blended with slag. *Iranian Journal of Materials Science and Engineering*, 2013. **10**(1): pp. 57–64.
8. Rashad, A.M., A comprehensive overview about the influence of different additives on the properties of alkali-activated slag–a guide for civil engineer. *Construction and Building Materials*, 2013. **47**: pp. 29–55.
9. Jiang, L. and D. Niu, Study of deterioration of concrete exposed to different types of sulfate solutions under drying-wetting cycles. *Construction and Building Materials*, 2016. **117**: pp. 88–98.
10. Mirza, J., et al., Preferred test methods to select suitable surface repair materials in severe climates. *Construction and Building Materials*, 2014. **50**: pp. 692–698.
11. Kumar, G.R. and U. Sharma, Abrasion resistance of concrete containing marginal aggregates. *Construction and Building Materials*, 2014. **66**: pp. 712–722.
12. Wang, B., S. Xu, and F. Liu, Evaluation of tensile bonding strength between UHTCC repair materials and concrete substrate. *Construction and Building Materials*, 2016. **112**: pp. 595–606.
13. Huseien, G.F., et al., Synthesis and characterization of self-healing mortar with modified strength. *Jurnal Teknologi*, 2015. **76**(1).
14. Sedaghatdoost, A., et al., Influence of recycled concrete aggregates on alkali-activated slag mortar exposed to elevated temperatures. *Journal of Building Engineering*, 2019. **26**: pp. 100871.
15. Alanazi, H., et al., Bond strength of PCC pavement repairs using metakaolin-based geopolymer mortar. *Cement and Concrete Composites*, 2016. **65**: pp. 75–82.
16. Huseien, G.F., K.W. Shah, and A.R.M. Sam, Sustainability of nanomaterials based self-healing concrete: an all-inclusive insight. *Journal of Building Engineering*, 2019.
17. Ouellet-Plamondon, C. and G. Habert, Life cycle assessment (LCA) of alkali-activated cements and concretes, in *Handbook of alkali-activated cements, mortars and concretes*. 2015: Elsevier. pp. 663–686.
18. McLellan, B.C., et al., Costs and carbon emissions for geopolymer pastes in comparison to ordinary portland cement. *Journal of Cleaner Production*, 2011. **19**(9–10): pp. 1080–1090.
19. Liu, Z., et al., Characteristics of alkali-activated lithium slag at early reaction age. *Journal of Materials in Civil Engineering*, 2019. **31**(12): p. 04019312.
20. Huseien, G.F., et al., Alkali-activated mortars blended with glass bottle waste nano powder: environmental benefit and sustainability. *Journal of Cleaner Production*, 2019: pp. 118636.
21. Wu, Y., et al., Geopolymer, green alkali activated cementitious material: synthesis, applications and challenges. *Construction and Building Materials*, 2019. **224**: pp. 930–949.
22. Shekhawat, P., G. Sharma, and R.M. Singh, Strength behavior of alkaline activated eggshell powder and flyash geopolymer cured at ambient temperature. *Construction and Building Materials*, 2019. **223**: pp. 1112–1122.
23. Behfarnia, K., *Studying the effect of freeze and thaw cycles on bond strength of concrete repair materials. Asian Journal of Civil Engineering*, 2010. **11**(2): pp. 165–172.

24. Gouny, F., et al., A geopolymer mortar for wood and earth structures. *Construction and Building Materials*, 2012. **36**: pp. 188–195.
25. Huseien, G.F., et al., Geopolymer mortars as sustainable repair material: a comprehensive review. *Renewable and Sustainable Energy Reviews*, 2017. **80**: pp. 54–74.
26. Chindaprasirt, P. and U. Rattanasak, Improvement of durability of cement pipe with high calcium fly ash geopolymer covering. *Construction and Building Materials*, 2016. **112**: pp. 956–961.
27. Hossain, M., et al., Durability of mortar and concrete containing alkali-activated binder with pozzolans: a review. *Construction and Building Materials*, 2015. **93**: pp. 95–109.
28. Norhasri, M.M., M. Hamidah, and A.M. Fadzil, Applications of using nano material in concrete: a review. *Construction and Building Materials*, 2017. **133**: pp. 91–97.
29. Pacheco-Torgal, F., et al., *Handbook of alkali-activated cements, mortars and concretes*. 2014: Elsevier.
30. Mirza, J., M. Mirza, and R. Lapointe, Laboratory and field performance of polymer-modified cement-based repair mortars in cold climates. *Construction and Building Materials*, 2002. **16**(6): pp. 365–374.
31. Emberson, N. and G. Mays, Significance of property mismatch in the patch repair of structural concrete Part 1: properties of repair systems. *Magazine of Concrete Research*, 1990. **42**(152): pp. 147–160.
32. Cusson, D. and N. Mailvaganam, Durability of repair materials. *Concrete International*, 1996. **18**(3): pp. 34–38.
33. Green, M.F., et al., Effect of freeze-thaw cycles on the bond durability between fibre reinforced polymer plate reinforcement and concrete. *Canadian Journal of Civil Engineering*, 2000. **27**(5): pp. 949–959.
34. Yang, Q., et al., Properties and applications of magnesia–phosphate cement mortar for rapid repair of concrete. *Cement and Concrete Research*, 2000. **30**(11): pp. 1807–1813.
35. Lee, M.-G., Y.-C. Wang, and C.-T. Chiu, A preliminary study of reactive powder concrete as a new repair material. *Construction and Building Materials*, 2007. **21**(1): pp. 182–189.
36. Lee, N., E. Kim, and H. Lee, Mechanical properties and setting characteristics of geopolymer mortar using styrene-butadiene (SB) latex. *Construction and Building Materials*, 2016. **113**: pp. 264–272.
37. Al-Zahrani, M., et al., Mechanical properties and durability characteristics of polymer- and cement-based repair materials. *Cement and Concrete Composites*, 2003. **25**(4): pp. 527–537.
38. Pane, I. and W. Hansen, Investigation of blended cement hydration by isothermal calorimetry and thermal analysis. *Cement and Concrete Research*, 2005. **35**(6): pp. 1155–1164.
39. Yang, Q. and X. Wu, Factors influencing properties of phosphate cement-based binder for rapid repair of concrete. *Cement and Concrete Research*, 1999. **29**(3): pp. 389–396.
40. Decter, M., Durable concrete repair—Importance of compatibility and low shrinkage. *Construction and Building Materials*, 1997. **11**(5): pp. 267–273.
41. Emberson, N. and G. Mays, Significance of property mismatch in the patch repair of structural concrete. Part 3: reinforced concrete members in flexure. *Magazine of Concrete Research*, 1996. **48**(174): pp. 45–57.
42. Emberson, N. and G. Mays, Significance of property mismatch in the patch repair of structural concrete Part 2: axially loaded reinforced concrete members. *Magazine of Concrete Research*, 1990. **42**(152): pp. 161–170.
43. Cabrera, J. and A. Al-Hasan, Performance properties of concrete repair materials. *Construction and Building Materials*, 1997. **11**(5): pp. 283–290.

44. Morgan, D., Compatibility of concrete repair materials and systems. *Construction and Building Materials*, 1996. **10**(1): pp. 57–67.
45. Pacheco-Torgal, F., et al., An overview on the potential of geopolymers for concrete infrastructure rehabilitation. *Construction and Building Materials*, 2012. **36**: pp. 1053–1058.
46. Davidovits, J., Geopolymer cement. A review. *Geopolymer Institute, Technical Papers*, 2013. **21**: pp. 1–11.
47. Pacheco-Torgal, F., J. Castro-Gomes, and S. Jalali, Alkali-activated binders: a review. Part 2. About materials and binders manufacture. *Construction and Building Materials*, 2008. **22**(7): pp. 1315–1322.

2 Geopolymer as Emerging Repair Materials

2.1 INTRODUCTION

The term "geopolymers" (GPs) was first coined by Joseph Davidovits in 1972 [1] to depict zeolite types of polymers. GPs that are being commonly prepared by the activation of slag, calcined clay (CA), fly ash (FA) and other aluminosilicate (AS) substances through an alkaline solution medium are established to be potential unconventional binding agents. GPs are the polymers of AS comprising three-dimensional disordered (glassy) structures which are formed due to the geopolymerization of AS monomers in the presence of alkali activation [2]. In the past, intensive studies on calcined clays (CAs) including metakaolin (MK) or various wastes from industries, such as FA, palm oil fuel ash (POFA) and slag, were performed [3–5]. Yet, the complex process, the so-called geopolymerization, is not completely known [6]. Davidovits proposed a reaction mechanism concerning the polycondensation of orthosilicate ions (kind of imaginary monomer) [7]. The mechanism of geopolymerization process [8] is based on the following stages: (i) alkali solution–assisted dissolution, (ii) restructuring and diffusion of dissolute ions and subsequent generation of tiny gelatinous organizations and (iii) formation of hydrated substances by the polycondensation of dissolved substances.

Compared to Ordinary Portland Cement (OPC), GPs are well known for their excellent properties, such as elevated compressive strength [9,10], little shrinkage [10,11], fire and acid resistant [12], devoid of emitting poisonous gases [13], poor thermal conduction [10], good immobilization of heavy metals, steadiness at high temperature [6], low energy production for building construction and several other industrial applications [10]. Owing to these distinctive features, GPs are potentially being employed in structure engineering, flame retardants, biomaterials and waste management [6,14]. New applications including the use of GP as a concrete repair material are under in-depth exploration.

In recent times, the use of GPs as surface concrete repair materials has generated renewed research interests [15–17]. In the exploitation of the GPs as new repairing substance, the binding efficacy among the concrete materials and the repairing substances [18,19] plays a decisive role. Geopolymer mortar (GPM) is compatible with Portland cement concrete because of the various similar properties including the elastic constants, tensile strength and the Poisson's ratio [20]. Furthermore, GPMs have the ability to repair at ambient temperature as traditional concretes [21]. These notable characteristics of GPs make them prospective towards surface concrete repair. Despite much research, the durability of GPs has not been assessed comprehensively for repair purposes.

Despite expensiveness, the commercial repair materials are generally employed for repairing concretes because of their excellent strength and binding capacity [22].

DOI: 10.1201/9781003173618-2

Thus, cheaper repairing substances with similar properties as substitute materials were demanded. Constant research efforts are made [23–26] to utilize GPs as repair materials, where tests are performed to determine their slant and direct shear, pull-out, and bonding strength of mortars and GPMs. Interestingly, GP exhibits elevated bond strength compared to OPC mixture. Torgal et al. [24] evaluated the bonding strength among concretes and GPMs that were made from the wastes of tungsten mine which contained calcium hydroxide (CaOH). It was revealed that GP binders possessed excellent bonding strength still at an early age as than the commercially available repairing materials. Songpiriyakij et al. [25] estimated the bonding strength of concretes and rebar using GP pastes (GPPs) as binding substance. It was acknowledged that the bonding strength of rice husk ash (RHA) and silica fume (SF) combined with GPs is roughly 1.5 times greater compared to epoxies. Consequently, the bond strength of GP materials is sufficiently high making it suitable as an alternative bonding material for repairing.

It is worth noting that millions of tons of natural, industrial and agriculture wastes, such as FA, coal and oil-burning by-products, bottom ash, POFA, RHA, bagasse ash (BA), used tyres, dust of cement, marble and crushed stone, waste ceramic materials and SF, are dumped every year in Malaysia. These waste materials cause severe ecological setbacks such as air contamination and leach out of hazardous substances. Several studies revealed that many of these wastes may be potentially recycled in the form of innovative concrete materials as an alternative to OPC (often as much as 70%). Besides, these newly developed concretes owing to their green chemical nature are environmentally friendly, durable and inexpensive building materials. Yet, the development of different GPs as repair materials especially for deteriorated concrete surfaces by containing the previously mentioned wastes is rarely explored.

As mentioned earlier, geopolymerization is a complex and important process in the GP industry, where high-PH alkaline solution which is used to dissolute the ASs is still to be clarified. The term "alkali activator" is used for a combination of a silica-rich solutions (e.g. sodium or potassium silicate) and highly concentrated alkali solutions (e.g. sodium or potassium hydroxide) with certain weight ratios. Such combination is used to dissolve alumina silicate from pozzolanic waste materials for building the amorphous structure of GPs. An increase in the ratio of silica-rich solution to alkali solution enhances the possibility of geopolymerization because of high amount of SiO_2. For various AS sources, it has been authenticated that the availability of SiO_2 is a key factor to determine the mechanism of geopolymerization.

Even though the knowledge regarding the mechanisms that control the alkali activation process is considerably advanced, many things need further investigations. The investigation on alkali-activated AS GPMs is comparatively a recent research domain than conventional OPC-based concretes. Activation of alkaline solution in ASs is distinct from the hydrated OPCs in terms of upper preliminary alkalinity and free of lime. These hydrated substances are distinct from several other products. Thus, depending on the chemical pathways of OPC-based products, it is very difficult to make proper predictions about the characteristics of ASs that are activated by alkaline solutions. The major binding stage of hydrated OPC is an aluminate-substituted calcium–silicate–hydrate (C-(A)-S-H) gel. Conversely, the foremost outcome obtained by the activation of alkaline solution is the sodium AS hydrate gel called

N-A-S-H. Thus, it is significant to determine the detailed mechanism of N-A-S-H formation.

Recent research indicated that calcium contents of FA affect significantly the resultant hardening characteristic of the GP where most of the earlier studies revealed promising results [27]. Calcium oxide (CaO) is assumed to generate calcium–silicate–hydrate (C-S-H) together with the AS GP gel. Major challenges for diverse applications of AS-based GPs are the need for high-temperature curing. Earlier researches were focused to increase the reactive nature of these substances through the incorporation of some Ca-based materials [28]. The incorporation of CaO allowed the formation of C-S-H gel together with AS GP networks. The contents of CaO in the precursor substance played a significant role to achieve the final hardening of GPs. Meanwhile, an increasing CaO content caused the enhancement in the mechanical characteristics and subsequent reduction in the time of setting.

The compatible nature of C-(A)-S-H and N-A-S-H gels has a significant influence on the hybrid OPC and alkaline solution–activated AS systems, wherein both products may be obtained [29]. Earlier researches used the synthetic gels to determine the influences of high pH levels on each gel's components on another. The aqueous aluminate is found to affect greatly the C-S-H product formation [30]. Besides, the aqueous Ca modified the N-A-S-H gels and partially replaced the sodium (Na) with Ca to produce (N, C)-A-S-H gels [30] as explained using the following reactions. However, the mechanisms for the formation of such gels and subsequent improvements are not completely understood. Furthermore, to explore the feasibility of achieving cements for the construction purposes, both gels must coexist and a methodical investigation on the compatible nature of N-A-S-H–C-A-S-H gel is essential.

2.2 GEOPOLYMER MORTAR

About four decades ago, Davidovits first coined the word *geopolymer* in the perspective of alkaline solution–activated MK [31]. Generally, GPs are regarded as synthetic inorganic polymer obtained by the activation of alkaline solution in ASs [32]. In recent years, GPs have gained renewed interest due to their environmentally affable attribute as an alternative to OPC [33]. In this view, the phenomenon called "geopolymerization" is a highly multifaceted exothermic process that involves many dissolution, reorientation and solidification reactions similar to zeolite production. In this process, high alkali component is employed for inducing the reaction among Si and Al elements. Actually, this concentrated alkaline solution is necessary to dissolve the ingredients and to produce polymer network structures in three dimensions which comprise interconnected chains of -Si-O-Al-O- bonding like:

$$M_n [- (SiO_2)_z - AlO_2]_n \cdot wH_2O$$

where M is the specific alkali atom (cation), including K, Na or Ca, the (-) symbols signify the appearances of chemical bonds, n refers to the degree of polycondensation or polymerization and index z can take value 1, 2, 3 or higher. The precise reaction pathways for explaining the setting time and hardening of GPs are not completely known so far. Moreover, it is believed that the geopolymerization process is decided by the presence of the so-called base material ASs and the actual compositions of the alkali solution used for activation [34]. Equation (2.1) below describes the process of geopolymerization that produces the polymer network consisting of Al and Si elements as backbone of 3D structures. According to Davidovits [35], the mechanism of geopolymerization engrosses the dissolution of ASs to create an Si–Al–based GP backbone as a short-ranged structure.

$$\text{Si-Al source} + \text{Water} + \text{Alkaline liquid} \rightarrow \text{GP precursor}$$

$$\text{GP precursor} + \text{Alkaline ions} \rightarrow \text{GP backbone} \tag{2.1}$$

Duxson et al. [32] explained the nature of the geopolymerization reactions in terms of the following three basic steps:

Step 1: Dissolution of the solid ASs: Being the pozzolanic solid, ASs are dissolved by the concentrated alkali activator via hydrolysis process at high-pH solution where a solution of silicate, aluminate and AS species is formed.

Step 2: Formation of gel: Species generated in the dissolution process are held in the aqueous phase, which also contains silicates appear from the activator solution. This supersaturated ASs solution produces a gel as oligomers to create long chains and networks. At this phase, water complexes that reside inside the pores are released.

Step 3: Polycondensation of ASs: The produced gels keep on rearranging continuously to form progressively larger network which finally creates 3D AS network of the GP binder.

Often, these progressions follow perpetually in the entire mixture in a non-linear time span. The dense inorganic GP binder material in the form of polymeric network thereby renders physiochemical properties superior to OPC. The following processes occur during the geopolymerization process:

- The complex network provides high compressive strengths to GPs [32,36].
- The distinct microstructures of the reaction products impart very good chemical resistance towards the deterioration of GP especially against sulphate, acid and seawater attacks [37].
- The GP matrix achieves high resistance as much as 1000°C–1200°C against thermal and fire assault [36].
- GPs often display swift setting without any prolonged degradation in the strength values [13,32].

The extent of dissolution of ASs in concentrated (high-pH) alkali solutions principally depends on the size distributions, morphologies and compositions of the source material particles [38–40], especially for glassy/disordered ASs. Earlier researches revealed that ASs-based GPs lack proper reactive element which resulted in deficient dissolution and subsequently achieved low mechanical strength [41]. When FA is used as the source of ASs for preparing GPs, precise characterization of the source material is necessary because the physical and chemical features of FA vary diversely and can affect the GP product formation [42–44]. Typical particle size of FA is found to be in the range of 1–150 μm [45] in the crystalline structures including quartz, mullite and various iron-rich phases, namely hematite [46]. The desired characteristics of resource substances for the synthesis of GPs thus depend on their final implementations. For the implementations of GPs, high-temperature–specific resource materials are needed to achieve desired mechanical performance and strengths.

GPs render a desired substitute to OPC binders. This is not only advantageous for the environment in the perspective of lessening the CO_2 emissions involving OPC production but also in terms of their durability performance. These properties are not only comparable but often superior to the one attained by OPC concrete. GP represents an alternative to PC due to similar or even better binding properties [47,48]. Recently, GP has attracted considerable attention due to its early compressive strength, low permeability, good chemical resistance and excellent fire resistance behaviour [4,32]. The suitability of many waste materials from industrial such as pulverized fuel ash (PFA), FA and POFA to produce geopolymer supported the prospect of GP binder to be an alternative to traditional OPC binder [49]. Compared to OPC concrete, the geopolymer concrete (GPC) yields very high performance in terms of mechanical strengths and durability [50–52].

It is found that ground-granulated blast furnace slag (GBFS) as waste material from the iron manufacturing industry has an important contribution in the production of high-strength GP concrete [53,54]. Malaysia and Thailand produce massive amounts of palm oil, where POFA is the waste material derived from burning the empty fruit, in generating electricity. Land filling using this POFA is a major environmental concern of this industry in these countries. Recently, POFA is realized as one of the most significant resources as binder material in GP production [55–57]. Other by-products from the industries such as RHA from the rice mills, red mud (RM) from the alumina refineries, copper and hematite mine tailings from mines [52,58–60] are also potential candidates for the production of GPMs.

2.3 GEOPOLYMER MORTAR AS REPAIR MATERIALS

Chronologically, the concept of GPs was first introduced as a new material by Joseph Davidovits in 1972 [35,61]. Those three-dimensional aluminium silicate inorganic polymers composed of AlO_4 and SiO_4 tetrahedral ions are mainly prepared from industrial wastes [62–66]. Their unique three-dimensional oxide network structure originating from inorganic polycondensation makes them advantageous. They possess several interesting features such as high strength, corrosion resistance, water resistance, high temperature resistance, and enclosed metal ions. [67,68]. GPs find broad range of applications in the field of transportation, emergency repairs, metallurgy, coating, membrane materials and nuclear waste disposal [69–74]. Despite significant commercial and technological potential, easy-brittle character of GPs limits their extensive applications [68].

Numerous studies are dedicated to optimize the strength of GP product and to understand the mechanism of geopolymerization. Bernal et al. [75] examined the development of binder structure in Na_2SiO_3-activated GBFS-MK blends to evaluate the effects of MK inclusion on the strength of resultant binder. Silva and Sagoe-Crenstil [76] inspected the Al_2O_3 to SiO_2 ratio dependent setting and hardening of the GP blends. This ratio has affected the setting time and the ultimate strength of GPs. Chindaprasirt et al. [77] investigated the influence of SiO_2:Al_2O_3 and Na_2O:SiO_2 ratios on the setting time, workability and the ultimate strength of GP specimens. It was acknowledged that the best ratio for GP binder is in the range of 2.87–4.79 for SiO_2/Al_2O_3 and within 1.2–1.4 for SiO_2/Na_2O. Bernal and Provis [78] addressed the durability of alkali-activated materials in terms of their recent progress and perspectives. The quick degradation tests were conducted to determine the effects of increasing contents of CO_2, sulphates and chlorides on the properties of GPs.

Over the years, GPs have been exploited as protective coating materials for marine concrete and transport systems [15–17]. However, the bonding strength among the concrete substrate and the repairing system [19,79,80] decides the binding efficiency of GPs as repairing material. The properties of GP concrete [20] including the elastic modulus, Poisson's ratio and tensile strength were analogous to the ordinary Portland cement (OPC) concrete. This clearly displayed the compatible nature of GP and POC concrete. Furthermore, alike conventional concrete the GP concrete can be cured at ambient temperature [21,81–83]. The degree of degradation of the GP concrete when soaked in the acid solution is significantly lower than that of Portland cement concrete [84,85]. Moreover, they possess low permeability and excellent anticorrosion property beneficial for efficient bond formation with cement paste and mortar [86]. They can be applied using the same apparatus and carry outs utilized for POC concrete for repairing degraded structures including pipes, manholes and chambers [87]. GPs' high-temperature stability makes them superior substitute to epoxy resins [88]. More significantly, manufacturing of FA-based GP cements emits 80%–90% lower CO_2 than OPC [89]. These advantages make GP an exceptional nominee as repairing materials. Despite much effort in the synthesis and characterization of GPs, their durability as repair material is far from being understood.

Generally, the repair work in concrete is widely performed using available repairing materials that have good compressive and bond strength [22] although expensive. Thus, less costly substitute repairing materials with analogous attributes are sought. Several researchers [23–25,90] attempted to employ GP as a repairing system due to their good slant and direct shear as well as pull-out. Hu et al. [23] investigated the bonding strength of the mortar substrate with GP systems where GP revealed greater bond strength than POC mixture. Pacheco-Torgal et al. [24] evaluated the bonding strength of concrete substrate with GPM prepared using wastes from tungsten mine comprising $Ca(OH)_2$. It was observed that GPs binders possessed great bonding strength even at an early age as compared to commercial repairing materials.

2.4 GEOPOLYMER PERFORMANCE

Several deteriorated concrete surfaces worldwide are increasingly facing problems related to the repairing or rehabilitation. Repairs of concrete structures are essential to ensure their service lifetimes, whereas they must be finished soon for public convenience. Many repairing materials are developed for concrete structures, such as cement-based materials, polymers and latex. These repair materials are commercialized and are diversely used. However, when cement-based materials are used for repairing massive concrete structures, considerable temperature cracks appear due to the generation of high hydration heat at the early age of curing. Furthermore, their use is limited because of the difficulty to achieve the required early age strength with appropriate workability. Other commercial materials such as polymers, polymer-modified materials and epoxy resin display good mechanical behaviour. However, they show very low durability in terms of low resistance to elevated temperatures and are very costly. In short, GP being one of the important materials is highly prospective for using as an alternative to existing repair materials.

It is demonstrated that GPs have great potential when utilized as repairing materials because they can achieve a high early strength and have a rapid setting time through an alkaline reaction in the presence of high activator contents. FA, GBFS and MK as waste are the best AS materials which are more environmentally sustainable than conventional mortar and commercial repair materials. GBFS being a high-calcium waste material, when blended with other waste materials (MK, FA and POFA) reveals many benefits in the GP industry. These advantages of GPMs include the reduced setting time, improved workability, curing at ambient temperature, high early strength and reduced porosity. By combining GBFS with FA, it is possible to enhance the resistance towards acid and sulphate attack together with the durability of GPMs. Blended POFA and GBFS-based GPMs are found to enhance the durability of GPMs to elevated temperatures.

The low $SiO_2:Al_2O_3$ of 2.1 and calcium content of FA class make it unsuitable to stand with AS materials for the preparation of GPMs [76,91]. It is reported that the optimum strength is ranged from 3 to 3.8. Calcium content less than 5% had great effect on the setting time and early strength of GPMs cured at ambient temperatures. It makes the system unsuitable for repairing work. For POFA waste material, the high ratio of $SiO_2:Al_2O_3$ of 15.1 also makes it unsuitable for preparing GPMs. Blended POFA with FA has been reported and about 30% of POFA replaced with FA contributed to achieve the highest strength and increase the resistance of mortar to elevated temperature, acid and sulphate attack [92–94]. Ceramic waste is one of the useful materials (rich AS) with the ratio of Si to Al around 5.9. Ceramic waste powder is proven to improve the concrete properties. Yet, no study has evaluated the effect of ceramic on mechanical and durability properties of GPs. Moreover, no investigation is made to assess the properties of ternary blend from the materials with high Ca content such as GBFS, high Al content such as FA class F and high silicate content such as POFA or ceramic waste powder. Table 2.1 shows the previous studies that used GP as a repair material.

It is authenticated that alkaline activator solution is an important factor in geopolymerization process of dissolution of the AS ions for forming the N-A-S-H network. Most of the previous researches have shown that the high molarity of NaOH (12–14 M) and the ratio of sodium silicate to sodium hydroxide around 2.5 are optimum for GP performance. Presently, one of the challenges is the high cost of GP that limits its wide range of construction applications. Besides, the high cost of alkaline solution, hazards of high molarity of NaOH and the severe environmental concerns are the major drawbacks of GPMs. Thus, rooms are open to produce environmentally friendly GP at commercial scale to fulfil the demand of construction industries.

TABLE 2.1
Overview of Previous Works on GP Mortars as Repair Materials

Ref.	Results
[27]	i. The GPM with high NaOH concentration containing PC as additive material revealed good performance in the shear bond strength prism test and bending stress of PCC notched beam test. The highest shear bond strength of 24.2 MPa was obtained with 14 M NaOH GP with 10% Portland cement. The bending stresses of PCC notched beams with filled GPM were enhanced as expected. The GPM mix with 14 M NaOH and 10% PC exhibited excellent bending stress of 3.1 MPa. However, with high NaOH concentration (14 M) and high PC (15%), a slight decrease in shear bond strength and bending stress was observed. The performance of GPM is found to be comparable to those of the commercial repair materials. ii. The average shear bond strength of RM is 17.9 MPa, while that of GPM was slightly higher (20.0 MPa). The average improvement of bending stresses of PCC notched beam with filled GPM and RM was 44% and 36%, respectively. Besides, the interface zones between PCC substrate and GPM containing PC as additive are dense and homogenous at contact zone due to the increase in reaction products. The results confirmed high bending stress and high *shear* bond strength of mixtures with a high amount of both NaOH and PC. This indicated the suitability of GPM containing PC as an additive to be applied as repair material.
[95]	i. The effect of sodium hydroxide and sodium silicate solutions as liquid portions on the mixture of FA–GBFS GP was investigated. The FA paste contained amorphous N-A-S-H gel and some crystalline phases of remaining FA. The increase in GBFS content enhanced the compressive strength and microstructure of GPP due to the formation of additional C-S-H. The use of NH and NHNS solutions resulted in crystalline C-S-H and amorphous gel, whereas the use of NS solution produced mainly the amorphous products. ii. For the FA and FA + GBFS pastes, the use of NH solution or NS solution alone gave low strengths when cured at ambient temperature. Better strength development is achieved through NHNS solution. For the GBFS paste, the presence of silicate enhanced the strength. Thus, the pastes containing NS solution performed better. Relatively high (28-day) compressive strengths of 171.7 and 173.0 MPa are obtained for GBFS pastes with NHNS and NS solutions, respectively. iii. The shear bond strength (slant angle of 45°) between concrete substrate and GPP is increased when the compressive strength and the amount of N-A-S-H gel of GPP are increased. The highest (28-day) shear bond strength of 31.0 MPa is achieved with FA + GBFS paste with NHNS solution. This indicates that it may be possible to use FA–GBFS GPP as a repair material. However, additional tests are required to confirm this observation.
[86]	i. MK/FA-based GPs cured at room temperature showed slightly lower bond strength than epoxy resin at ambient temperature. Yet, they retained much higher bond strength throughout the temperature range from 100°C to 300°C. ii. The bond strength of pure MK-based GPs revealed significant degradation in the range of 20°C–100°C due to the evaporation of free water and crack propagation. iii. GPs with low Si/Al ratio and high FA/(MK + FA) ratio exhibited lower bond strength at ambient temperature. However, it retained higher bond strength at elevated temperatures. iv. Bond strength of GPs is increased slightly with a decrease in SiO_2/K_2O ratio. v. Too low or too high solid-to-liquid ratio is not beneficial to enhance the bond strength of GPs. For achieving good strength and optimum workability conditions in MK/FA-based GPs, the optimum solid-to-liquid ratio is in the range of 0.6–0.8.

(Continued)

TABLE 2.1 (Continued)
Overview of Previous Works on GP Mortars as Repair Materials

Ref.	Results
[96]	i. OPA appears more advantageous than PWA, as a supplementary raw material in GPs due to their better overall strength characteristics. ii. Both OPA and PWA dramatically increased the bond strength to Portland cement mortar and may be necessary components in GPs to be used for concrete repair. iii. Compressive strength of GPs can be improved by increased heat curing (up to 4 hours). However, only 10% of OPA with heat curing at 80°C for 1 hour showed maximal strength. Meanwhile, with 10% PWA (longer cure times) affected early strength development. iv. Long heat curing times also reduced drying shrinkage, potentially due to the well-developed strength.
[97]	i. One of the largest disadvantages of geopolymeric binders is that they are more expensive than those based on Portland cement. This high cost is essentially due to the expensive chemical activators. It is concluded that MK-GPM with low sand/binder mass ratio present low adhesion to concrete substrate due to high shrinkage behaviour deduced by the microcracks in the surface of specimens. Although the mortars tested showed adhesion strength lowers than the commercial repair mortars, the former is much more cost-effective (5–10 times less expensive).
[98]	i. The results revealed that the mortar workability is reduced with the increase in sodium hydroxide concentration and high replacement of MK with calcium hydroxide. This is because MK has a high Blaine fineness. It is observed that the compressive strength and flexural strength are enhanced with the increase in sodium hydroxide concentration (35%) in both cases. The combination of super-plasticizer (3%) and calcium hydroxide (10%) enhanced mortar flow from less than 50% to over 90% while maintaining a high level of mechanical strength. The use of super-plasticizer content up to 3% did not reduce the mechanical strength, except for the mixture with a calcium hydroxide content of 10% and a sodium hydroxide concentration of 12 M.
[99]	i. The compressive strengths of geopolymeric green cement using 30 wt% FA and 70 wt% GBFS as raw materials can be reached to 47 MPa when prepared with 0.96 SiO_2/Na_2O molar ratio alkali solutions. ii. In building repairing test, repair rates are up to 120% and 110% for cement mortar specimen tension tests and shear tests, respectively. It verified that geopolymeric green cement is a good adhesive bonding material.
[23]	i. The compressive strength of the cement repair material is found to be lower than that of geopolymeric repair material with or without steel slag at 8 hours, 1, 3, 7, and 28 days. ii. The geopolymeric repair materials have better repair characteristics than cement-based repair materials. The bond strength of GSb at 3 d is discerned to be 2.6%, which is 600% higher than those of Gb and Cb alone, respectively. Similarly, the bond strength of GSb at 28 d is 4.4%, which is 55.9% higher than those of Gb, and Cb alone, respectively. iii. The geopolymeric repair materials possess better abrasion resistance than cement repair. The PG values at 3, 7, 28, 56 and 90 days are decreased to 48%, 44%, 29%, 28% and 29% than PC, respectively. The addition of steel slag improved significantly the abrasion resistance performance of the geopolymeric repair. Comparing PGS with PG, the p-values at 3, 7, 28, 56 and 90 d are found to decrease at 9%, 8.9%, 21.4%, 22.2% and 22.5%, respectively. The steel slag is almost fully absorbed to take part in the alkali-activated reaction and immobilized into the amorphous alumina silicate GP matrix as revealed by SEM analyses.

(*Continued*)

TABLE 2.1 (Continued)
Overview of Previous Works on GP Mortars as Repair Materials

Ref.	Results
[100]	i. Tungsten mine waste geopolymeric binder possessed much higher bond strength than current commercial repair products. That advantage is higher at early ages. ii. Commercial repair products gained no bond to sawn concrete specimens. Geopolymeric type achieved the highest strength with substrate surface. iii. SEM micrographs displayed that the tungsten mine waste geopolymeric binder gets chemically bonded to the concrete substrate. iv. Cost comparisons between tungsten mine waste geopolymeric binder and current commercial repair products showed that geopolymeric ones are by far the most cost-efficient solution.

Low NaOH molarity and low amount of sodium silicate are the important factors that reduce the GPM cost, health and environmental problems of existing GP. Materials with high calcium content such as GBFS showed the ability of Ca^{++} to replace part of Na^+ in geopolymerization process. This in turn produces C-S-H gel and C-A-S-H gel besides the N-A-S-H gel and improves the performance of GP. On high NaOH molarity, the dissolution of Ca is suppressed, resulting in less hydration products [95].

In brief, an all-inclusive review of the existing literature revealed that the binary blend of GBFS and FA or POFA activated with high-concentration alkaline solution is beneficial in improving the performance of GP. Most of the researchers are focused to evaluate the mechanical properties of GP. They depended on the compressive and bond strength as critical factors to evaluate GPM as a repair material for repairing the concrete surface deterioration. There is a lack of information on the durability of GP as a repair material such as abrasion resistance, freezing–thawing, bond strength in an aggressive environment and elevated temperatures.

2.5 DURABILITY AND SUSTAINABILITY

According to Nuaklong et al. [101], the GP made from high Ca content FA revealed low chloride penetration depth and high sulphuric acid resistance. Khankhaje et al. [102] studied POFA-based mortar which showed a satisfactory durability in the laboratory. Ariffin et al. [93] and Bhutta et al. [103] examined GP (BAG) concrete blends by adding PFA and POFA with alkali solution activators in the presence of 2% of H_2SO_4 and 5% of Na_2SO_4 solutions cured for 18 months. It was shown that GPC are more resistant to acid attack than OPC concrete due to the total removal of cement in the blend. After 18 months of exposure to 2% of H_2SO_4 solution, GPC containing POFA and PFA suffered from a mass loss of 8% which was much below the mass loss (20%) of OPC concrete. Furthermore, GPC displayed a 35% loss in the strength, while the loss of OPC concrete was 68% after 18 months.

Bhutta et al. [103] declared that there was no considerable damage to the surface of GPC after the exposure to 5% Na_2SO_4 solution for 1 year. Hussin et al. [94] investigated the effects of high temperatures on the characteristics of GPC with POFA and PFA. It was concluded that GPC acquired an improved structural stability than OPC concrete after exposure to 800°C. However, GPC exhibited the cracks at the surface in the temperature range of 600°C–800°C. Conversely, OPC concrete developed cracks at much lower temperature of 200°C. Bakharev [104,105] showed that GPCs are greatly resistant to sulphate and acid attack. Rajamane et al. [106] showed that the penetration of chloride ion in GPC and conventional concrete (CC) is very low. Sathia et al. [107] investigated the absorption properties and acid attack resistance of GPC and shown that GPCs have good durability.

Ganesan et al. [108] reported the properties of GP including water absorption, efficient porosity, and sorptivity. The sorptivity of GPC was lower than that of CC, while the abrasion resistance (AR) was higher than that of CC. The mean wear resistance of GPC was 27.5% lower than that of CC, and for steel fibre reinforced geopolymer concrete (SFRGPC), it was almost 65% lower than that of CC specimens. GPC products showed excellent resistance to acid and sulphate attack and suffered less than 2% weight loss when exposed to 3% H_2SO_4 solution for 6 months. Corresponding weight loss for CC specimens was 27%. Both GPC and CC specimens suffered less than 1% weight loss when subjected to sulphate attack. Mathew et al. [109] reported that GP is a sustainable construction material. In fact, sodium hydroxide (39%) and sodium silicate (49%) together contributed a lion's share to embodied energy of GPC. Moreover, in OPC cement it contributed nearly 94% of the total embodied energy.

2.6 ECONOMIC FEASIBILITY

Mathew et al. [109] showed that the estimated cost of GPC is double compared to OPC-based concrete. Studies should be done in the area of manufacturing process of sodium hydroxide to reduce the cost. Figure 2.1 shows the fine aggregate cost minimization using alternate material such as crusher dust. However, the impact of using such materials on the strength of concrete needs careful examination.

Dai, et al. [99] reported that the cost for the preparation of GPM is about 2,900$ per ton. Although it is slightly higher than that of OPC, it has decided advantages in terms of ambient temperature processing, low carbon dioxide emission, environmental friendliness, carbon dioxide reduction target and reutilization of the waste. Torgal et al. [100] evaluated the mechanical performance of commercial repair materials and found them similar to GPM. Current commercial repair materials are very

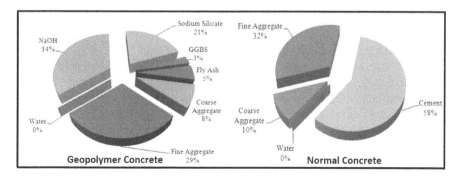

FIGURE 2.1 Cost of each material to FA-GBFS-based GPC and OPC-based concrete [110].

TABLE 2.2
Flexural Strength and Cost-to-Strength Ratio (Economic Analysis) [100]

Item	CRM1	CRM2	CRM3	CRM4	CRM5	GPM1	GPM2	GPM3
Flexural strength, MPa	0.80	1.30	1.80	0.30	0	3	4.1	7.1
Cost-to-strength ratio, Euros/MPa	2,275	1,400	1,011	10,973	0	91	66	38

expensive at least in the early ages. Repairing concrete substrate with tungsten mine waste GPM is 11 times cheaper as summarized in Table 2.2.

2.7 ENVIRONMENT SUITABILITY AND SAFETY FEATURES

During the process of OPC concrete, it emits much CO_2 as a reaction product which arises from the de-carbonation of lime and calcination of cement clinker. Some CO_2 is also regenerated due to the utilization of alkaline solution hydroxide or silicate as an activator rather during cement hydration process. Production of these activators requires temperature similar to de-carbonation of lime in OPC manufacturing. The CO_2 emission of GPM can be estimated in terms of the compositions. Study found that 110 kg of activator was required to be mixed with 400 kg pozzolan to produce 1 m^3 of GPM which emits CO_2 equal to 27.5% of the same amount of OPC. When pozzolan is utilized in the natural state, the production of 1 ton of OPC released about 1 ton of CO_2. However, in calcined form the CO_2 emission of alkali activated natural pozzolan (AANP) concrete is the summation of CO_2 emission because it generates the required activators and the amount is connected to the calcination process. The temperature required for calcination of these materials is half of that needed to de-carbonate the lime. Thus, the CO_2 liberation for calcinations of these materials can be regarded as 50% to that of OPC manufacturing. The CO_2 emission of GPC is enhanced to 77.5% of the amount emitted by the same quantity of OPC. Therefore, the GPC manufacturing was accountable to reduce the CO_2 emission from 22.5% to 72.5% than the OPC production [111].

A world without concrete is quite unimaginable. In fact, without concrete it would be impossible for magnificent buildings including the Sydney Opera House, the Chrysler Building, or Taj Mahal to exist. The skyscrapers of cities would have no way to reach such impressive heights if they were not made of concrete. Moreover, without the durability of concrete, historic buildings would have withered away centuries ago. Certainly, concrete is necessary for everyday life in its own right. In short, the production of modern concrete carries with it a heavy price. The manufacturing of concrete causes billions of tons of raw materials to be wasted every year due to inefficiency of concrete production. Additionally, the production of Portland cement, the main binder in concrete, contributes to more than 5% of the total greenhouse gases released globally each year. This poses risks in a world where sustainable and green building have become a major issue. Thus, the future dream would be to create a cleaner, efficient, reliable and even stronger substitute to concrete than those currently used. The obvious answer is GPC [110].

2.8 MERITS AND DEMERITS OF GEOPOLYMER AS REPAIR MATERIAL

Bondar and Dali [111] reported the benefits of mortar made with alkaline solution–activated alumina silicate than other binders in terms of carbon footprint and cost. Increased pressure to improve sustainability within the concrete industry makes them very significant. The correlation between CO_2 footprint and cost of GPM and its compositions in comparison with OPC-based cements is roughly quantified. GP shows high durability compared with OPC and other commercial repair materials, and excellent properties within both acid and salt environments [93,94,103–105]. The GPM revealed high early mechanical properties than other materials [97–100,112,113]. GP is superior to Portland cement for several reasons such as far lower carbon footprint, less cracking, more resistant to corrosive elements such as sea salt, excellent frost resistance and durability in cold climates with rapid set binders available. The high cost of GPC binders is one of the major factors which remains a severe shortcoming over OPC [114–116]. Currently, GP binders only become economically competitive for high-performance structural purposes. In short, the above-cited disadvantage means that the study of GP applications should focus on high-cost materials such as commercial concrete repair mortars.

Torgal et al. [24] showed that GPMs are nearly seven times cheaper than the current commercially available repairing mortars. But if the cost-to-bond strength ratio is compared, the differences are even higher, with the cost of the cheapest commercial repair mortar being 13.8 times higher than the GPMs. The GPMs present a stiff workability behaviour arising from the use of viscous compounds such has sodium silicate and sodium hydroxide. Numerous reports hinted the placement difficulties related to the low workability of GPMs. It is displayed [117] that several super-plasticizers used in the Portland cement concrete industry lost their fluidity for GPMs. Other authors [118] found out that the use of a super-plasticizer leads to an improvement of GPMs' workability, but they can also contribute to a reduction in compressive strength depending on sodium silicate to sodium hydroxide ratio. Sathonsaowaphak et al. [119] reported that workable ranges of sodium silicate/sodium hydroxide ratios and sodium hydroxide concentration are between 0.67–1.5 and 7.5–12.5 M, respectively. Also, Rangan [120] confirmed that the addition of a naphthalenesulfonate-based super-plasticizer improved the workability of FA GP mixtures. However, super-plasticizer content above 2% is responsible for a slight degradation of compressive strength.

2.9 SUMMARY

The following conclusions were drawn:

i. Traditional cement-based materials are used to repairing massive concrete structures; considerable temperature cracks appear due to the generation of high hydration heat at the early age of curing. Furthermore, their use is limited because of the difficulty to achieve the required early age strength with appropriate workability.

ii. Polymers, polymer-modified materials and epoxy resin display good mechanical behaviour. However, they show very low durability in terms of low resistance to elevated temperatures and at very high cost.
iii. It is demonstrated that GPs have great potential when utilized as repairing material because they can achieve a high early strength and have a rapid setting time through an alkaline reaction in the presence of high activator contents.
iv. FA, GBFS and MK as waste are the best AS materials which are used as calcium resources in GP preparation.
v. GP being one of the important materials is highly prospective for using as an alternative to existing repair materials.

REFERENCES

1. Komnitsas, K. and D. Zaharaki, Geopolymerisation: a review and prospects for the minerals industry. *Minerals Engineering*, 2007. **20**(14): pp. 1261–1277.
2. Rowles, M. and B. O'Connor, Chemical optimisation of the compressive strength of aluminosilicate geopolymers synthesised by sodium silicate activation of metakaolinite. *Journal of Materials Chemistry*, 2003. **13**(5): pp. 1161–1165.
3. Chang, J., A study on the setting characteristics of sodium silicate-activated slag pastes. *Cement and Concrete Research*, 2003. **33**(7): pp. 1005–1011.
4. Kong, D.L., J.G. Sanjayan, and K. Sagoe-Crentsil, Comparative performance of geopolymers made with metakaolin and fly ash after exposure to elevated temperatures. *Cement and Concrete Research*, 2007. **37**(12): pp. 1583–1589.
5. Temuujin, J., A. van Riessen, and K. MacKenzie, Preparation and characterisation of fly ash based geopolymer mortars. *Construction and Building Materials*, 2010. **24**(10): pp. 1906–1910.
6. Yao, X., et al., Geopolymerization process of alkali–metakaolinite characterized by isothermal calorimetry. *Thermochimica Acta*, 2009. **493**(1): pp. 49–54.
7. Provis, J., et al., The role of mathematical modelling and gel chemistry in advancing geopolymer technology. *Chemical Engineering Research and Design*, 2005. **83**(7): pp. 853–860.
8. Dimas, D., I. Giannopoulou, and D. Panias, Polymerization in sodium silicate solutions: a fundamental process in geopolymerization technology. *Journal of Materials Science*, 2009. **44**(14): pp. 3719–3730.
9. Burciaga-Díaz, O., R. Magallanes-Rivera, and J. Escalante-García, Alkali-activated slag-metakaolin pastes: strength, structural, and microstructural characterization. *Journal of Sustainable Cement-Based Materials*, 2013. **2**(2): pp. 111–127.
10. Zhang, Y.J., Y.C. Wang, and S. Li, Mechanical performance and hydration mechanism of geopolymer composite reinforced by resin. *Materials Science and Engineering: A*, 2010. **527**(24): pp. 6574–6580.
11. Chi, M.-c., J.-j. Chang, and R. Huang, Strength and drying shrinkage of alkali-activated slag paste and mortar. *Advances in Civil Engineering*, 2012. **2012**: pp. 1–7.
12. Palomo, A., et al., Chemical stability of cementitious materials based on metakaolin. *Cement and Concrete Research*, 1999. **29**(7): pp. 997–1004.
13. Duxson, P., et al., The role of inorganic polymer technology in the development of 'green concrete'. *Cement and Concrete Research*, 2007. **37**(12): pp. 1590–1597.
14. Davidovits, J. 30 years of successes and failures in geopolymer applications. Market trends and potential breakthroughs. in *Keynote conference on geopolymer conference*. 2002: Melbourne, Australia. pp. 1–16.

15. Balaguru, P., *Geopolymer for protective coating of transportation infrastructures.* Center for Advanced Infrastructure and Transportation (CAIT), 1998: pp. 1–29.
16. Zhang, Z., X. Yao, and H. Wang, Potential application of geopolymers as protection coatings for marine concrete III. Field experiment. *Applied Clay Science*, 2012. **67**: pp. 57–60.
17. Zhang, Z., X. Yao, and H. Zhu, Potential application of geopolymers as protection coatings for marine concrete: II. Microstructure and anticorrosion mechanism. *Applied Clay Science*, 2010. **49**(1): pp. 7–12.
18. Geissert, D.G., et al., Splitting prism test method to evaluate concrete-to-concrete bond strength. *ACI Materials Journal*, 1999. **96**: pp. 359–366.
19. Momayez, A., et al., Comparison of methods for evaluating bond strength between concrete substrate and repair materials. *Cement and Concrete Research*, 2005. **35**(4): pp. 748–757.
20. Hardjito, D., et al. Introducing fly ash-based geopolymer concrete: manufacture and engineering properties. in *30th conference on our world in concrete and structures.* Singapore, 2005: pp. 1–8.
21. Nath, P. and P.K. Sarker, Use of OPC to improve setting and early strength properties of low calcium fly ash geopolymer concrete cured at room temperature. *Cement and Concrete Composites*, 2015. **55**: pp. 205–214.
22. Mirza, J., et al., Preferred test methods to select suitable surface repair materials in severe climates. *Construction and Building Materials*, 2014. **50**: pp. 692–698.
23. Hu, S., et al., Bonding and abrasion resistance of geopolymeric repair material made with steel slag. *Cement and concrete composites*, 2008. **30**(3): pp. 239–244.
24. Pacheco-Torgal, F., J. Castro-Gomes, and S. Jalali, Adhesion characterization of tungsten mine waste geopolymeric binder. Influence of OPC concrete substrate surface treatment. *Construction and Building Materials*, 2008. **22**(3): pp. 154–161.
25. Suksiripattanapong, C., et al., Compressive strength development in fly ash geopolymer masonry units manufactured from water treatment sludge. *Construction and Building Materials*, 2015. **82**: pp. 20–30.
26. Huseien, G.F., et al., Effects of ceramic tile powder waste on properties of self-compacted alkali-activated concrete. *Construction and Building Materials*, 2020. **236**: p. 117574.
27. Phoo-ngernkham, T., et al., High calcium fly ash geopolymer mortar containing Portland cement for use as repair material. *Construction and Building Materials*, 2015. **98**: pp. 482–488.
28. Al-Majidi, M.H., et al., Development of geopolymer mortar under ambient temperature for in situ applications. *Construction and Building Materials*, 2016. **120**: pp. 198–211.
29. Yip, C.K., G. Lukey, and J. Van Deventer, The coexistence of geopolymeric gel and calcium silicate hydrate at the early stage of alkaline activation. *Cement and Concrete Research*, 2005. **35**(9): pp. 1688–1697.
30. Garcia-Lodeiro, I., et al., Compatibility studies between NASH and CASH gels. Study in the ternary diagram Na_2O–CaO–Al_2O_3–SiO_2–H_2O. *Cement and Concrete Research*, 2011. **41**(9): pp. 923–931.
31. Davidovits, J., *Geopolymer chemistry and applications.* 2008: Geopolymer Institute, ISBN 978-2-9514820-1-2.
32. Duxson, P., et al., Geopolymer technology: the current state of the art. *Journal of Materials Science*, 2007. **42**(9): pp. 2917–2933.
33. Rickard, W.D., et al., Assessing the suitability of three Australian fly ashes as an aluminosilicate source for geopolymers in high temperature applications. *Materials Science and Engineering: A*, 2011. **528**(9): pp. 3390–3397.
34. Ghosh, K. and P. Ghosh, Effect Of Na_2O/Al_2O_3, SiO_2/Al_2O_3 and w/b ratio on setting time and workability of fly ash based geopolymer. *International Journal of Engineering Research and Applications*, 2012. **2**(4): pp. 2142–2147.

35. Davidovits, J., *Geopolymers. Journal of Thermal Analysis*, 1991. **37**(8): pp. 1633–1656.
36. Andini, S., et al., Coal fly ash as raw material for the manufacture of geopolymer-based products. *Waste Management*, 2008. **28**(2): pp. 416–423.
37. Provis, J.L., Geopolymers and other alkali activated materials: why, how, and what? *Materials and Structures*, 2014. **47**(1): pp. 11–25.
38. Chen-Tan, N.W., et al., Determining the reactivity of a fly ash for production of geopolymer. *Journal of the American Ceramic Society*, 2009. **92**(4): pp. 881–887.
39. Fernández-Jiménez, A., A. Palomo, and M. Criado, Microstructure development of alkali-activated fly ash cement: a descriptive model. *Cement and Concrete Research*, 2005. **35**(6): pp. 1204–1209.
40. Bakharev, T., Thermal behaviour of geopolymers prepared using class F fly ash and elevated temperature curing. *Cement and Concrete Research*, 2006. **36**(6): pp. 1134–1147.
41. Rahier, H., J. Denayer, and B. Van Mele, Low-temperature synthesized aluminosilicate glasses Part IV Modulated DSC study on the effect of particle size of metakaolinite on the production of inorganic polymer glasses. *Journal of Materials Science*, 2003. **38**(14): pp. 3131–3136.
42. Manz, O.E., Coal fly ash: a retrospective and future look. *Fuel*, 1999. **78**(2): pp. 133–136.
43. McCarthy, G.J., et al., *Fly ash and coal conversion by-products: characterization, utilization and disposal III.* 1987: Materials Research Society.
44. Day, R. and F. Glasser, *Fly ash and coal conversion by-products: characterization, utilization and disposal 6.* 1990: Materials Research Society.
45. Berry, E. and V.M. Malhotra. Fly ash for use in concrete-a critical review. in *Journal proceedings. ACI*, 1980. **77**(2): pp. 59–73.
46. Matsunaga, T., et al., Crystallinity and selected properties of fly ash particles. *Materials Science and Engineering: A*, 2002. **325**(1): pp. 333–343.
47. Shi, C., A.F. Jiménez, and A. Palomo, New cements for the 21st century: the pursuit of an alternative to Portland cement. *Cement and Concrete Research*, 2011. **41**(7): pp. 750–763.
48. Pacheco-Torgal, F., J. Castro-Gomes, and S. Jalali, Alkali-activated binders: a review. Part 2. About materials and binders manufacture. *Construction and Building Materials*, 2008. **22**(7): pp. 1315–1322.
49. Joseph, B. and G. Mathew, Influence of aggregate content on the behavior of fly ash based geopolymer concrete. *Scientia Iranica*, 2012. **19**(5): pp. 1188–1194.
50. Görhan, G. and G. Kürklü, The influence of the NaOH solution on the properties of the fly ash-based geopolymer mortar cured at different temperatures. *Composites Part B: Engineering*, 2014. **58**: pp. 371–377.
51. Giasuddin, H.M., J.G. Sanjayan, and P. Ranjith, Strength of geopolymer cured in saline water in ambient conditions. *Fuel*, 2013. **107**: pp. 34–39.
52. Nazari, A., A. Bagheri, and S. Riahi, Properties of geopolymer with seeded fly ash and rice husk bark ash. *Materials Science and Engineering: A*, 2011. **528**(24): pp. 7395–7401.
53. Aydın, S. and B. Baradan, Mechanical and microstructural properties of heat cured alkali-activated slag mortars. *Materials & Design*, 2012. **35**: pp. 374–383.
54. Islam, A., et al., The development of compressive strength of ground granulated blast furnace slag-palm oil fuel ash-fly ash based geopolymer mortar. *Materials & Design*, 2014. **56**: pp. 833–841.
55. Yusuf, M.O., et al., Strength and microstructure of alkali-activated binary blended binder containing palm oil fuel ash and ground blast-furnace slag. *Construction and Building Materials*, 2014. **52**: pp. 504–510.
56. Mijarsh, M., M.M. Johari, and Z. Ahmad, Synthesis of geopolymer from large amounts of treated palm oil fuel ash: application of the Taguchi method in investigating the main parameters affecting compressive strength. *Construction and Building Materials*, 2014. **52**: pp. 473–481.

57. Ban, C.C., P.W. Ken, and M. Ramli, Effect of sodium silicate and curing regime on properties of load bearing geopolymer mortar block. *Journal of Materials in Civil Engineering*, 2016: p. 04016237.
58. He, J., et al., Synthesis and characterization of red mud and rice husk ash-based geopolymer composites. *Cement and Concrete Composites*, 2013. **37**: pp. 108–118.
59. Ahmari, S. and L. Zhang, Production of eco-friendly bricks from copper mine tailings through geopolymerization. *Construction and Building Materials*, 2012. **29**: pp. 323–331.
60. Chen, Y., et al., Preparation of eco-friendly construction bricks from hematite tailings. *Construction and Building Materials*, 2011. **25**(4): pp. 2107–2111.
61. Gordon, L.E., J.L. Provis, and J.S. van Deventer, Non-traditional ("geopolymer") cements and concretes for construction of large CCS equipment. *Energy Procedia*, 2011. **4**: pp. 2058–2065.
62. Shvarzman, A., et al., The effect of dehydroxylation/amorphization degree on pozzolanic activity of kaolinite. *Cement and Concrete Research*, 2003. **33**(3): pp. 405–416.
63. ul Haq, E., S.K. Padmanabhan, and A. Licciulli, Synthesis and characteristics of fly ash and bottom ash based geopolymers–a comparative study. *Ceramics International*, 2014. **40**(2): pp. 2965–2971.
64. Usha, S., D.G. Nair, and S. Vishnudas, Geopolymer binder from industrial wastes: a review. *International Journal of Civil Engineering and Technology*, 2014. **5**(12): pp. 219–225.
65. Ibrahim, W., et al. Development of fly ash-based geopolymer lightweight bricks using foaming agent-a review. in *Key engineering materials*. 2015: Trans Tech Publ.
66. Hardjito, D., et al. Brief review of development of geopolymer concrete. in *Invited paper, George Hoff symposium*. 2004: American Concrete Institute.
67. Singh, B., et al., Geopolymer concrete: a review of some recent developments. *Construction and Building Materials*, 2015. **85**: pp. 78–90.
68. Zhang, S., K. Gong, and J. Lu, Novel modification method for inorganic geopolymer by using water soluble organic polymers. *Materials Letters*, 2004. **58**(7): pp. 1292–1296.
69. Van Jaarsveld, J. and J. Van Deventer, The effect of metal contaminants on the formation and properties of waste-based geopolymers. *Cement and Concrete Research*, 1999. **29**(8): pp. 1189–1200.
70. Xu, H. and J. Van Deventer, The geopolymerisation of alumino-silicate minerals. *International Journal of Mineral Processing*, 2000. **59**(3): pp. 247–266.
71. Van Deventer, J., et al., Reaction mechanisms in the geopolymeric conversion of inorganic waste to useful products. *Journal of Hazardous Materials*, 2007. **139**(3): pp. 506–513.
72. He, Y., et al., Preparation of self-supporting NaA zeolite membranes using geopolymers. *Journal of membrane science*, 2013. **447**: pp. 66–72.
73. Zhang, J., et al., Synthesis of a self-supporting faujasite zeolite membrane using geopolymer gel for separation of alcohol/water mixture. *Materials Letters*, 2014. **116**: pp. 167–170.
74. Zhang, Z., et al., Preparation and characterization of a reflective and heat insulative coating based on geopolymers. *Energy and Buildings*, 2015. **87**: pp. 220–225.
75. Bernal, S.A., et al., Evolution of binder structure in sodium silicate-activated slag-metakaolin blends. *Cement and Concrete Composites*, 2011. **33**(1): pp. 46–54.
76. De Silva, P. and K. Sagoe-Crenstil, The effect of Al_2O_3 and SiO_2 on setting and hardening of $Na_2O-Al_2O_3-SiO_2-H_2O$ geopolymer systems. *Journal of the Australian Ceramic Society*, 2008. **44**(1): pp. 39–46.
77. Chindaprasirt, P., et al., Effect of SiO_2 and $Al2O_3$ on the setting and hardening of high calcium fly ash-based geopolymer systems. *Journal of Materials Science*, 2012. **47**(12): pp. 4876–4883.
78. Bernal, S.A. and J.L. Provis, Durability of alkali-activated materials: progress and perspectives. *Journal of the American Ceramic Society*, 2014. **97**(4): pp. 997–1008.

79. Geissert, D.G., et al., Splitting prism test method to evaluate concrete-to-concrete bond strength. *ACI Materials Journal*, 1999. **96**(3): pp. 359–366.
80. Momayez, A., Comparison of methods for evaluating bond strength between concrete substrate and repair materials. *Cement and Concrete Research*, 2005. **35**(4): pp. 748–757..
81. Nath, P. and P.K. Sarker, Effect of GGBFS on setting, workability and early strength properties of fly ash geopolymer concrete cured in ambient condition. *Construction and Building Materials*, 2014. **66**: pp. 163–171.
82. Deb, P.S., P. Nath, and P.K. Sarker, The effects of ground granulated blast-furnace slag blending with fly ash and activator content on the workability and strength properties of geopolymer concrete cured at ambient temperature. *Materials & Design*, 2014. **62**: pp. 32–39.
83. Deb, P.S., P. Nath, and P.K. Sarker. Properties of fly ash and slag blended geopolymer concrete cured at ambient temperature. in *7th international structural engineering and construction conference*. 2013: Honolulu, USA.
84. Bakharev, T., J. Sanjayan, and Y.-B. Cheng, Resistance of alkali-activated slag concrete to acid attack. *Cement and Concrete Research*, 2003. **33**(10): pp. 1607–1611.
85. Rangan, B.V., et al. Studies on fly ash-based geopolymer concrete. in *Proceedings of the world congress geopolymer*. 2005: Saint Quentin, France.
86. Zhang, H.Y., et al., Characterizing the bond strength of geopolymers at ambient and elevated temperatures. *Cement and Concrete Composites*, 2015. **58**: pp. 40–49.
87. Montes, C. and E. Allouche, Evaluation of the potential of geopolymer mortar in the rehabilitation of buried infrastructure. *Structure and Infrastructure Engineering*, 2012. **8**(1): pp. 89–98.
88. Pacheco-Torgal, F., et al., An overview on the potential of geopolymers for concrete infrastructure rehabilitation. *Construction and Building Materials*, 2012. **36**: pp. 1053–1058.
89. Davidovits, J. Properties of geopolymer cements. in *First international conference on alkaline cements and concretes*. 1994: Kiev State Technical University, Ukraine.
90. Songpiriyakij, S., et al., Anchorage of steel bars in concrete by geopolymer paste. *Materials & Design*, 2011. **32**(5): pp. 3021–3028.
91. Duxson, P., et al., Understanding the relationship between geopolymer composition, microstructure and mechanical properties. *Colloids and Surfaces A: Physicochemical and Engineering Aspects*, 2005. **269**(1): pp. 47–58.
92. Ariffin, M., et al. Mix design and compressive strength of geopolymer concrete containing blended ash from agro-industrial wastes. in *Advanced Materials Research*. 2011: Trans Tech Publ.
93. Ariffin, M., et al., Sulfuric acid resistance of blended ash geopolymer concrete. *Construction and Building Materials*, 2013. **43**: pp. 80–86.
94. Hussin, M., et al., Performance of blended ash geopolymer concrete at elevated temperatures. *Materials and Structures*, 2015. **48**(3): pp. 709–720.
95. Phoo-ngernkham, T., et al., Effects of sodium hydroxide and sodium silicate solutions on compressive and shear bond strengths of FA–GBFS geopolymer. *Construction and Building Materials*, 2015. **91**: pp. 1–8.
96. Hawa, A., et al., Development and performance evaluation of very high early strength geopolymer for rapid road repair. *Advances in Materials Science and Engineering*, 2013. **2013**: pp. 1–10.
97. Vasconcelos, E., et al. Concrete retrofitting using CFRP and geopolymer mortars. in *Materials science forum*. 2013: Trans Tech Publ.
98. Moura, D., et al. Concrete repair with geopolymeric mortars: influence of mortars composition on their workability and mechanical strength. in *VI international materials symposium (materials 2011), Portugal* . 2011: pp. 1–6.
99. Dai, Y., et al., *A study on application of geopolymeric green cement*. National Ilan University. 2013: pp. 1–8.

100. Torgal, F.P., J. Gomes, and S. Jalali, Bond strength between concrete substance and repair materials: comparisons between tungsten mine waste geopolymeric binder versus current commercial repair products. *Seventh International Congress on Advances in Civil Engineering, Yildiz Technical University*, 2006: pp. 1–10.
101. Nuaklong, P., V. Sata, and P. Chindaprasirt, Influence of recycled aggregate on fly ash geopolymer concrete properties. *Journal of Cleaner Production*, 2016. **112**: pp. 2300–2307.
102. Khankhaje, E., et al., On blended cement and geopolymer concretes containing palm oil fuel ash. *Materials & Design*, 2016. **89**: pp. 385–398.
103. Bhutta, M.A.R., et al., Sulfate and sulfuric acid resistance of geopolymer mortars using waste blended ash. *Jurnal Teknologi*, 2013. **61**(3).
104. Bakharev, T., Resistance of geopolymer materials to acid attack. *Cement and Concrete Research*, 2005. **35**(4): pp. 658–670.
105. Bakharev, T., Durability of geopolymer materials in sodium and magnesium sulfate solutions. *Cement and Concrete Research*, 2005. **35**(6): pp. 1233–1246.
106. Rajamane, N., et al., Rapid chloride permeability test on geopolymer and Portland cement concretes. *Indian Concrete Journal*, 2011. **85**(10): p. 21.
107. Sathia, R., K.G. Babu, and M. Santhanam. Durability study of low calcium fly ash geopolymer concrete. in *Proceedings of the 3rd ACF international conference-ACF/VCA, Vietnam*. 2008: pp. 1–7.
108. Ganesan, N., R. Abraham, and S.D. Raj, Durability characteristics of steel fibre reinforced geopolymer concrete. *Construction and Building Materials*, 2015. **93**: pp. 471–476.
109. Mathew, M.B.J., M.M. Sudhakar, and D.C. Natarajan, Strength, economic and sustainability characteristics of coal ash–GGBS based geopolymer concrete. *International Journal of Computational Engineering Research*, 2013. **3**(1): pp. 207–212.
110. Faridmehr, I., M.L. Nehdi, M. Nikoo, G.F. Huseien, and T. Ozbakkaloglu, Life-cycle assessment of alkali-activated materials incorporating industrial by-products. *Materials*, 2021. **14**(9): p. 2401..
111. Bondar, D., *Alkali activation of Iranian natural pozzolans for producing geopolymer cement and concrete*. 2009: The University of Sheffield.
112. Phoo-ngernkham, T., et al., The effect of adding nano-SiO_2 and nano-Al_2O_3 on properties of high calcium fly ash geopolymer cured at ambient temperature. *Materials & Design*, 2014. **55**: pp. 58–65.
113. Phoo-ngernkham, T., et al., Properties of high calcium fly ash geopolymer pastes with Portland cement as an additive. *International Journal of Minerals, Metallurgy, and Materials*, 2013. **20**(2): pp. 214–220.
114. Davidovits, J., Chemistry of geopolymeric systems, terminology. in *Proceedings of 99 international conference*. eds. Joseph Davidovits, R. Davidovits & C. James, 1999": France.
115. Deventer, S. *Opportunities ans obstacles in the commercialisation of geopolymers*. 2002: Geopolymer.
116. Harper, R., W. South, and R. Knigth. Geopolymers–a commercial reality. in *Proceedings of 2002 geopolymer conference*. 2002: Melbourne, Australia.
117. Palacios, M. and F. Puertas, Effect of superplasticizer and shrinkage-reducing admixtures on alkali-activated slag pastes and mortars. *Cement and Concrete Research*, 2005. **35**(7): pp. 1358–1367.
118. Chindaprasirt, P., T. Chareerat, and V. Sirivivatnanon, Workability and strength of coarse high calcium fly ash geopolymer. *Cement and Concrete Composites*, 2007. **29**(3): pp. 224–229.
119. Sathonsaowaphak, A., P. Chindaprasirt, and K. Pimraksa, Workability and strength of lignite bottom ash geopolymer mortar. *Journal of Hazardous Materials*, 2009. **168**(1): pp. 44–50.
120. Provis, J.L. and J.S.J. Van Deventer, *Geopolymers: structures, processing, properties and industrial applications*. 2009: Elsevier.

3 Manufacturing Geopolymer
Materials and Mix Design

3.1 INTRODUCTION

Rapid deterioration of concrete surface being a major problem in the construction materials industry requires further improvement. It is well known that these concrete structures get deteriorated progressively due to various reasons related to physical, chemical, thermal and biological processes. Over the years, numerous expensive surface repair mortars are developed and commercially available on the marketplace for repairing and maintenance. Lately, geopolymer mortars (GPMs) as alternative repair materials have received focused attention. These types of materials are tested in terms of the parameters influencing the behaviour of the product, reaction mechanisms and mechanical characteristics, particularly the compressive strength and bonding between conventional concrete and GPM.

It is worth noting that millions of tons of natural, industrial and agriculture wastes, such as fly ash (FA), coal and oil-burning by-products, bottom ash, palm oil fuel ash (POFA), bagasse ash (BA), used tyres, dust from cement, marble and crushed stone, and waste ceramic materials, are dumped every year in Malaysia. These waste materials cause severe ecological setbacks such as air contamination and leach out of hazardous substances. Several studies [1–4] revealed that many of these wastes may be potentially recycled in the form of innovative concrete materials as an alternative to Ordinary Portland Cement (OPC) (often as much as 70%). Besides, these newly developed concretes owing to their green chemical nature are environmentally friendly, durable and inexpensive building materials. Recent research indicated that calcium contents of FA affect significantly the resultant hardening characteristic of the geopolymers (GPs) where most of the earlier studies revealed promising results [5,6]. Calcium oxide (CaO) is assumed to generate calcium–silicate–hydrate (C-S-H) together with the N-A-S-H gel. The major challenges for diverse applications of N-A-S-H-based GPs are the need for high-temperature curing. Earlier researches were focused to increase the reactive nature of these substances through the incorporation of some calcium-based materials [7,8]. The incorporation of CaO allowed the formation of C-S-H gel together with N-A-S-H networks [5,9]. The contents of CaO in the precursor substance played a significant role to achieve the final hardening of GPs. Meanwhile, an increasing CaO content caused the enhancement in the mechanical characteristics and subsequent reduction in setting time [7,10]. Palomo et al. [11] developed two models to understand the binding characteristics of GPs with alkaline solution activation. The first model concerns with the mild alkaline solution activation of silica (Si) plus Ca substances including ground-granulated blast furnace slag (GBFS) to produce C-A-S-H gels as the main product (called alkali-activated mortars). The second model deals with the alkaline solution activation of Si plus Al substances including FA that needs a robust alkali solution to produce N-A-S-H gels as the major outcome (called GPMs).

GPs can be prepared by activating several waste substances comprising high contents of aluminosilicates (ASs), including FA [12–14], slag [15,16] and POFA [17–20], and can be synthesized and produced using high alkaline concentration. During the last two decades, there have been a number of researchers who focused their efforts to utilize GPM due to the wide range of potential applications [21–23]. The effects of single and binary combination of binders among metakaolin (MK), GBFS and POFA have been reported [24–26]. This chapter provides a comprehensive literature survey on the most important agriculture, construction and by-product wastes that can be used to manufacture GPs. The effect of these materials on workability, mechanical and durable performance of proposed GP was discussed.

3.2 FLY ASH-BASED GEOPOLYMER BINDER

FA is one of the most abundant materials on the Earth. It is also a crucial component for the production of GPMs because of its role in the geopolymerization process. Being a pozzolanic material, it exhibits cementitious properties when combined with calcium hydroxide (CaOH). FA is the main by-product created from the combustion of coal in coal-fired power plants which are classified as Class C and Class F types based on their CaO contents. Class C FA has a high Ca content, which is mainly generated from the burning of lignite coal sources. This class has total SiO_2, Al_2O_3 and Fe_2O_3 contents between 50 wt.% and 70 wt.% and CaO content more than 20 wt.%. Class F FA has a low Ca content and is generated from burning anthracite or bituminous coal. It has total SiO_2, Al_2O_3 and Fe_2O_3 contents over 70 wt.% and CaO content less than 10% [27,28]. Each class of such FA has its own unique properties, and both of them are suitable for GP industry.

In recent times, the use of FA is attracting a lot of interest of researchers because of its excellent performance, lower cost and carbon dioxide emission footprint. As a waste material, FA is relatively cheaper than other materials. The associated CO_2 emission is comparably low and performs better at elevated temperature. Less sodium silicate solution is required to activate FA- and GBFS-based materials, thereby making their environmental impact lower than others [29]. In GPs, FA is the source of the ASs for the binder and is thus the critical component for strength development. FA is a powdery material made up of small glass spheres, consisting primarily of Si, Al, FE and CaO [30]. Fernandez-Jimenez et al. [31] acknowledged one limitation of FA-based GP for wide application is the requirement of curing at elevated temperature. Bakharev [32,33] observed that FA being a low amorphous content with relatively large average particle size produces low-strength GPs. Literature revealed that GP source material must be carefully selected and accurately characterized to optimize its effectiveness in producing usable GPs.

3.2.1 Effect of FA on Workability and Strength Properties

The properties of FA-based GP depend on the chemical constituents and the amount of FA as well as activator solution properties and curing method. It is known that, percentage of Na$_2$O (by weight of FA) and SiO$_2$:Na$_2$O ratio of the mix considerably influence the workability and setting time of GP where the former one depends on the viscosity. The viscosity of the gel rises with time due to the process of geopolymerization. Okoye et al. [34] reported that in the binary blend GP contenting FA and silica fume workability shows increment with an increase in FA content in GP mixture. Duan et al. [35] studied the workability of GP content sawdust and FA which was increased with increasing FA content. Chindaprasirt et al. [36] reported that the fineness of FA significantly affects the setting time of GP, whereas an increase in the fineness of FA reduces the setting time of GP. Furthermore, it is observed that the flow of GP mortars is enhanced with an increase in FA fineness. Hardjito et al. [37] examined the effect of increasing FA content on the setting time of GPMs. The curing condition is found to affect the setting time of FA-based GPM. It is also observed that the samples which are cured at high temperature of 65°C–90°C presented short setting compared to those cured at room temperature.

The properties of FA-based GP have been studied in the last decades [38–40]. They are favourable for cementitious mortar usage due to their excellent durability. Many researchers [41–44] reported similar engineering properties of GP mortar that are potential for construction. Despite intensive research, FA-based GPs are still far from practical applications on a large scale and many problems need to be overcome [45]. It is reported [46,47] that FA-based GP revealed the best strength when cured in the temperature range of 65°C–80°C.

3.2.2 Effect of FA on Durability of Geopolymer

The durability of FA-based GP including the resistance to chloride, sulphate, sulphuric acid, freeze–thaw, thermal and efflorescence is examined by many researchers. Durability is closely related to the microstructure and the migration behaviour of ions from FA-based GP. These in turn can be adjusted by the alkali solution, curing method and adding Ca during the preparation. FA provides excellent mechanical properties and good resistance to severe environments. Several studies are performed to evaluate the GP durability. Zahang et al. [48] reported high durability of FA-based GP for coating concrete materials in marine environment. They showed high-durability performance for thermal application [49,50]. Chindaprasirt et al. [51] reported that FA-based GP displays high resistance to acid attack in aggressive environments. Chindaprasirt et al. [51] showed that the durability of cement pipe is improved in the presence of FA-based GP as a covering material in aggressive solutions. Ariffin et al. [52] acknowledged that FA-based GP presents high resistance to H$_2$SO$_4$ compared to conventional mortar.

3.3 PALM OIL FUEL ASH

Huge quantities of the palm oil waste (POFA) are obtained from the agro industries in Malaysia, Indonesia and Thailand where POFA is the by-product of palm oil. In addition, although due to the increased plantation of palm oil trees, this production rate is likely to increase [53]. POFA is derived by burning empty fruit bunches, oil palm clinkers and shell for electricity generation. A survey revealed that the annual production of POFA in the year 2007 by Malaysia and Thailand was nearly 3 million tons and 100,000 tons, respectively. This tendency kept on increasing due to the plantation of more palm trees [54,55]. Interestingly, this material having no market value is simply dumped into the ponds/lagoons [56] as land fillers, which is a serious environmental concern. However, recent research [57,58] has paved the way for the development of sustainable material using POFA and related waste materials from the palm oil industry. Such advancement added further research impetus because it is identified that POFA is rich in silica. Yet, POFA is classified as a pozzolanic substance [33–36]. This is used for the partial substitution of OPC in the traditional concrete to enhance its strength and durability [53,56,59,60].

3.3.1 Effect of POFA on Workability and Strength Properties

Salami et al. [61] examined the effect of POFA on the workability of GPM. It is found that POFA mortar in fresh state has stiff consistency due to the resistance to flow offered by high surface area of POFA. Furthermore, the workability of POFA mortar is enhanced with increasing solution content of ($NaOH + Na_2SiO_3$) to binder. This is attributed to the total amount of water increment in the specimen. Salih et al. [62] have reported that the increase in the ratio of (Na_2SiO_3:$NaOH$) from 1.0 to 3.0 could reduce the workability of POFA mortar. This was mainly attributed to a higher content of water in lower ratio (1.0) than the higher one (3.0). The majority of the previous researches [43,44,52,63,64] revealed that the values of compressive strength varies from 28 to 66 MPa at low volume levels of POFA (almost 30%) which was mixed with slag, FA and rice husk ash under a curing temperature of 65°C and 75°C for the period of 24–48 hours. It was further shown that an increase in the POFA content more than 30% led to reduce the compressive strength of the mortar.

Salih et al. [20,62,65] also studied the effect of 100% POFA with comparatively high volume of Ca cured at 60°C for the period of 2 hours. Compressive strength value up to 32 MPa was achieved wherein POFA was considered as a high Ca–activated substance. The strength was linked to the generation of C-S-H gel from Ca and Si elements present in POFA and the participated silicates from the activator. Salih et al. [20] examined the influence of varying curing temperatures on the alkaline solution activation of POFA. It was demonstrated that the strength at early ages was enhanced at a higher rate when cured in the oven. Moreover, an increase in the curing temperature from 60°C to 80°C did not really influence the compressive strength. At late ages up to 90 days, mortars cured at ambient temperature revealed roughly the same strength than those cured at higher temperatures. This confirmed the effectiveness of curing at ambient temperature and subsequent geopolymerization of POFA as the AS source. Furthermore, the strength progress at ambient temperature for ages between 7 and 90 days was higher than those cured at oven temperature. It was acknowledged that 70°C was the optimum temperature for the activation of POFA with Na_2SiO_3 and NaOH. Besides, the strength was dropped for oven-cured (80°C) samples wherein further curing reduced the hardening time because of the increase in the rate of geopolymerization at higher temperatures. The micro-cracks were found to propagate with increasing curing time. Conversely, specimens cured at ambient temperature did not reveal any surface cracks.

Yusuf et al. [17,66] have shown a reduction in the flexural and compressive strengths of the 70% POFA blend specimens where the least strength was increased with Si/Al ratio, leading to the existence of reaction relics. The compressive strength is increased with decreasing MK replacement up to 50% and thereafter started reducing up to a measured POFA replacement of 80%. The compressive strengths for 80% POFA were discerned to be 26.17 and 30.791 MPa for 3 and 7 days, respectively. These strength values are remarkable for low-strength mortar. It was established that the replacement of POFA with MK also improved the workability and reduced the demand for water in the GPM.

3.3.2 Effect of POFA on Durability of Geopolymer

Alkali-activated binders have several notable attributes, such as high early strength, low creep, low shrinkage to good resistance against acid and sulphate in addition to its environment friendliness [67,68]. Salmi et al. [61] have shown an increase in the weight loss of GPM specimens with increasing NaOH molarity. The least and most weight loss of specimens exposed to 5% of $MgSO_4$ and 5% of Na_2SO_4 solutions, respectively, was recorded. GP containing POFA displayed a high resistance towards elevated temperatures, sulphate and acid attack [52,69].

3.4 GROUND BLAST FURNACE SLAG

Ground blast furnace slag (GBFS) is a waste material which is obtained by quenching molten iron slag (a by-product of iron and steel-making) from a blast furnace in water or steam. This produces a glassy and granular product which is then dried and grounded into a fine powder. The chemical composition of a slag varies considerably depending on the composition of the raw materials in the Fe production process. GBFS shows cementitious and pozzolanic properties due to the high content of CaO and SiO_2. Furthermore, GBFS has been widely used in construction industry to improve the durability of conventional concrete and enhance the mechanical properties [70], the microstructure and the durability of GPM [71].

3.4.1 Effect of GBFS on Workability and Strength Properties

Nath et al. [71] studied the influence of GBFS on the workability and setting time of FA-based GPM where GBFS replaced FA with 0, 10, 20 and 30 mass%. The flow of mortar was found to reduce for more than 50%. The increasing level of GBFS from 0% to 30% further reduced the initial and final setting time. Islam et al. [63] showed that the highest compressive strength is about 66 MPa for 70% GBFS with 30% POFA. However, the use of large amount of GBFS led to decline the workability and setting time. Sahana [72] revealed that an inclusion of GBFS at different replacement levels below 40% could increase the setting time of the mortar. However, beyond this level, the setting time could reduce and lead to a loss of workability and casting of concrete. Deb et al. [16] described a reduction in the workability of high GBFS content. This is due to the accelerated reaction of the Ca and the angular shape of GBFS. Though alkali-activated slag has high strength, the issues related to rapid setting and insufficient workability along with the high values of dry shrinkage limited their applicability [73].

Al-majidi et al. [7] studied the impact of increasing GBFS content on the FA-based GPM. The workability was decreased and the setting times (initial and final) as well as hardening were accelerated. At ambient curing condition (23°C), FA-based GPs blended with GBFS or OPC were discerned to reduce the setting time to a value comparable to that of OPC. The workability was measured in terms of the flow of mortars, which showed a slight decrease due to the presence of additives and faster rate of setting [74]. Al-Majidi et al. [7] have cured FA-based GPs at ambient temperature to determine their compressive strength. It considerably affected the blend composition and improved the compressive strength with increasing contents of GBFS to total binder ratio in GPM mixtures. Increasing GBFS content from 10% to 50% of the total binder enhanced the compressive strength from 18.45 to 48 MPa at 28 days. The effect of GBFS content on the development of flexural and direct tensile strengths in ambient temperature–cured GPM was similar to that observed in the development of compressive strength. GBFS content of 40% showed optimum flexural and tensile strength with 6 and 3 MPa, respectively. Nath et al. [74] studied the impact of GBFS or OPC inclusion on the compressive strength of FA-based GP. The compressive strength was increased with the increase in the binder content for FA blended with 10% GBFS or OPC.

Puertas et al. [75] examined the compressive strength and hydrated specimens of FA-combined GBFS pastes. The strength attained over 50 MPa at 28 days for the specimen containing FA: BGFS of 1.0 which was activated with 10 M NaOH solution and cured at 25°C. Shi and Day [76] showed that the strength of early-age curing of FA/slag blend activated with NaOH and Na_2SiO_3 was enhanced considerably due to the inclusion of a little amount of hydrated lime. Nath et al. [71] acknowledged that by adding slag up to 30% to the total binder it is possible to achieve a concrete strength of over 55 MPa and that of mortar as much as 63 MPa at 28 days. The compressive strength values of the mortars were enhanced with the addition of GBFS amount while the flexural strength values were decreased with the increasing rate of FA. The main reason for this was ascribed to the varying activation of different binders, the particle size and shape of the binders [77]. Earlier studies [10,21,78–80] indicated that an increasing level of GBFS in FA or POFA in the GP mixture could reduce the workability and setting time, and enhanced the mechanical performance.

3.4.2 Effect of GBFS on Durability of Geopolymer

Increasing level of GBFS in FA-based GPM mixture was found to improve the microstructure and enhanced the dense structure and lowered the porosity [7]. Li and Liu [81] reported that blended slag with FA-based GP could enhance the durability properties of GPM and showed good resistance to permeation. Blended GBFS with

FA-based GP can elevate temperature resistance [82] and sodium sulphate attack [83]. Moreover, it suffers deterioration in magnesium sulphate attack [83] and the shrinkage become higher [84]. The rate of deterioration was reported to decrease with increasing slag content in the FA–slag blend. Microscopic images of slag blended FA-based GP revealed mostly amorphous and Ca-containing hydration product. The compactness of the gel is increased when the slag content was higher in the mortar, thereby enhancing durability [71].

3.5 CERAMIC WASTES

The ceramic powder is the principal waste of the ceramic industry which is generated as unwanted dust during the process of dressing and polishing. It is estimated that 15%–30% of the ceramic wastes are produced from the total raw material used. A portion of this waste is often utilized on-site for the excavation pit refilling. Ceramic waste can be used in concrete to improve the strength and other durability factors. Fernandes et al. [85] reported that the waste contents at various fabrication phases in the ceramic industries can reach nearly 3%–7% of its global manufacturing. This specifies that huge amount of calcined clays is just dumped for landfilling every year. Moreover, the deposition processes are becoming expensive due to the ever-growing constraints on landfilling. In that case, industries must look for alternative solutions such as recycling such waste materials as useful products. Despite some reuses of ceramic wastes, the quantities of such wastes utilized by the construction sector are still negligible [86]. Thus, its immediate reuse in other industries appears essential. Building sector being the customer of much ceramic wastes will continue to play a vital role to overcome some of the environmental issues. The concrete industries can use the ceramic wastes safely without requiring any remarkable change in the production and application process. Moreover, the cost of deposition of ceramic waste in landfill can be saved together with the replacement of raw materials and natural resources thus saving energy and protecting the environment. Some studies suggested that the construction industry can be more sustainable and beneficial if most of the industrial wastes can be recycled effectively as useful GP concretes [87,88].

3.5.1 Effect of Ceramic Wastes on Workability and Strength Properties

Samadi et al. [89] demonstrated that the ceramic waste powder (CWP) has positive effect on the compressive strength when CWP is replaced by OPC with less than 40% where the microstructure properties of mortar are enhanced. Increase in the level of replacement by more than 40% led to reduce the compressive strength. Senthamarai et al. [90] reported that ceramic waste can be used as coarse aggregate in concrete. The basic trend of permeation characteristics of the ceramic waste coarse aggregate concrete is similar to those of the conventional concrete. Ariffin et al. [91] investigated the effect of ceramic that replaced sand as fine aggregate in GP mortars. The results revealed that the mortar prepared with ceramic aggregate presented high strength compared to conventional GPMs. The effect of CWP on fresh and hardened GPMs is far from being researched.

3.5.2 Effect of Ceramic Waste on Durability of Geopolymer

Pacheco et al. [86] showed that the concrete combined with CWP has increased durability performance because of its pozzolanic properties. It was realized that by replacing the conventional sand with CWP it is possible to achieve mortars with superior strength and durability performance. This CWP-substituted conventional coarse aggregate mortars are prospective but performed little low towards water absorption. Water permeability implies that the substitution of conventional sand by CWP is an excellent option. They did not examine the durability of CWP in terms of resistance against sulphuric acid and sulphate attack as well as elevated temperatures.

3.6 ALKALINE ACTIVATOR SOLUTIONS

Alkali-activated materials are generally classified into two groups. The first one includes the high-calcium system with GBFS as a typical precursor and C-A-S-H type gel as the main reaction product [92]. The second type is the low calcium system with Class F FA, POFA and MK as representative raw materials and N-A-S-H type gels within a three-dimensional network as the major reaction product [93]. Extensive research has been performed on these two systems to determine the role of activator type and alkali concentration [44], the effect of the dosage of raw materials [94], the effect of admixtures [95–97], the curing effect [98], microstructure, mechanical properties, thermal properties and durability [96]. In spite of excellent performances of both systems, there remains several shortcomings for practical applications such as fast setting, high shrinkage of alkali-activated slag [99,100], elevated curing temperature demand and relatively long setting times of alkali-activated ASs. Such disadvantages are overcome using a promising solution made of blended alkaline systems (Na_2O–CaO–Al_2O_3–SiO_2 systems) which are produced by mixing calcium-enriched precursors and ASs [39,43,63,101–104].

In geopolymerization, the activation of raw resource materials using alkaline solution is very effective to achieve the desired strength of mortars. A strong alkali activator is vital for the surface hydration enhancement of the ASs present in the raw material. In addition, the chemical activator contents affect significantly the mechanical strengths of GPMs [105,106]. Komljenovic et al. [107] showed that the types and amount of alkaline solution play a dominant role during activation. Alkali activators such as $Ca(OH)_2$, NaOH, NaOH + Na_2CO_3, KOH and Na_2SiO_3 with different contents were used to make FA-based GPMs where the curing condition was kept constant. The effects of these activators on the strength behaviour of the GPMs were examined. Results on the compressive strength revealed that Na_2SiO_3 has maximum activation followed by the others where KOH revealed lower activation than NaOH. This was attributed to the dissimilarity in the ionic radii of Na and K [106]. Irrespective of the alkali activator types, the compressive strength was enhanced with increasing activator contents. The estimated optimum value of modulus was found for Na_2SiO_3 value of 1.5 where higher modulus values than the recommended one caused adverse influence on the strength of GPMs. Conversely, the mixing of Si and Al elements during the GP preparation depends on the NaOH contents. Actually, the quantity and time of leaching for Si and Al is critically decided by the NaOH contents [108]. The dissolution, hydrolysis and condensation reaction of GPs are greatly affected by the effective Si:Al ratios. In low Si:Al ratio GP system, the condensation reaction tends to occur between aluminate and silicate species, resulting in mainly poly(sialate) geopolymeric structures. Moreover, the condensation reaction in high Si:Al system would predominantly create the silicate species itself. This forms the oligomeric silicates which in turn condense with $Al(OH_4)^{+4}$ and forms geopolymeric structures of poly(sialate-siloxo) and poly(sialate-disiloxo) [109,110].

3.6.1 WORKABILITY AND STRENGTH PERFORMANCE

Ghosh et al. [111] reported the effects of water content increase in the blend on the reduction of alkaline activator molar concentration. This in turn reduced the viscosity of mix and slowed down the rate of geopolymerization together with workability enhancement. A linear decrease in the flow diameter was observed with the percentage increase of Na_2O. Moreover, an increase in the percentage of Na_2O enhanced the viscosity of the blend and reduced the flow diameter. The flow diameter and viscosity of GP mix was reduced with increasing alkaline solution concentration. Furthermore, the flow diameter revealed a linear decrease with the increase in the percentage of SiO_2. The viscosity of GPM was enhanced with increasing dosage of soluble silicates.

Sathonsaowaphak et al. [112] examined the influence of different factors including NaOH contents, Na_2SiO_3 to NaOH ratio and alkali solution to binder ratio (S:B) on the workability and strength development of GPMs. Water and super-plasticizer were also added to improve the workability of the mortar mixes without changing the mortar strength. Results displayed that the ratios of S to B, Na_2SiO_3 to NaOH and NaOH contents were in the range of 0.42–0.71, 0.67–1.5 and 10 M, respectively. GPMs achieved good compressive strength with improved workability. It was asserted that the inclusion of 10 M NaOH solution was crucial for the geopolymerization as the Na^+ ions balanced the charges. On top of this, the dissolution rate of Si and Al was enhanced due to the presence of NaOH. However, POFA-based GPs, the optimum ratios of solid to liquid and Na_2SiO_3:NaOH achieved the highest compressive strength of 1.32 and 2.5, respectively [62]. Existence of higher void density in the ratio of solid to liquid below 1.32 has negatively affected the strength properties. Furthermore, the ratio of Na_2SiO_3 to NaOH more than 2.5 slowed down the rate of geopolymerization.

Sukmak et al. [113] studied the effect of sodium Na_2SiO_3 to NaOH and liquid alkaline activator to FA binder (S:B) ratios on the development of compressive strength for clay FA–based GP bricks under extended curing ages. The ratios of Na_2SiO_3:NaOH were 0.4, 0.7, 1.0, 1.5 and 2.3 and the S:B ratios were 0.4, 0.5, 0.6 and 0.7 by dry clay mass. The brick specimens were compacted using the manual hydraulic jack at the optimum water content to get the highest dry unit weight. Then, these brick specimens were left to set at room temperature for 24 hours before being oven-cured at 75°C for 48 hours. The compressive strength tests were conducted on 7, 14, 28, 60 and 90 days of curing ages. Results revealed that S:B ratios below 0.3 and above 0.8 are unsuitable for the synthesis of such bricks because the strength reduced to zero for these ratios. The best ratios of Na_2SiO_3:NaOH and S:B were discerned to be 0.7

and 0.6, respectively. The best Na_2SiO_3:NaOH ratio of 0.7 is below the FA-based GPs. The clay possessing high-cation absorption ability absorbed some of the added NaOH and reduced the strength significantly. This decrease for clay FA–based GP bricks with excessive alkali activator (S:B > 0.6) was ascribed to the precipitation of dissolved Si and Al elements at the early ages before starting the polycondensation. This caused the crack formation on the FA particles. The optimum compressive strength was observed to be nearly 15 MPa at 90 days of curing.

Ridtirud et al. [34] determined the optimal ration of 1.5 for Na_2SiO_3 to NaOH in FA-based GPMs. These GPMs with SS to SH ratios of 0.33, 0.67, 1.0, 1.5 and 3.0 displayed an increasing strength of 25.0, 28.0, 42.0, 45.0 and 23.0 MPa, respectively. This observation was majorly ascribed to the enhanced Na contents in the mixes where Na^+ ion played a significant role in the development of GP and acted as charge balancing entities. Moreover, too much silicate contents in the GP system could reduce the compressive strength by hampering the water evaporation and disrupting the formation of 3D networks of ASs in the mortar. Gorhan and Kurklu [114] studied the effects of varying NaOH solution concentrations (3, 6 and 9 M) on the 7-day compressive strength of Class F FA-based GPMs. Other parameters including GBFS:FA and Na_2SiO_3:NaOH were kept constant. The results revealed that the optimum NaOH concentration of 6 M produced the maximum compressive strength of 22.0 MPa at 7 days of curing. This alkaline concentration rendered an ideal environment for appropriate dissolution of FA particles without hindering the polycondensation. Very low NaOH contents (3 M) could not induce strong chemical reaction. However, very high NaOH content (9 M) caused premature coagulation of Si and manifested lower compressive strength of GPMs.

Somna et al. [115] investigated the compressive strength of FA cured at ambient temperature with changing NaOH concentration from 4.5 to 16.5 M. Increase in the NaOH contents from 4.5 to 9.5 M caused a considerable increase in the compressive strength of the paste. The changes of NaOH contents from 9.5 to 14 M also enhanced the compressive strength of the specimen. The increase in compressive strength with the increase in NaOH contents was majorly attributed to the high amount of Si and Al leaching. The decrease in compressive strength of FA-hardened pastes at the NaOH contents of 16.5 M was ascribed to the surplus hydroxide ions that caused the precipitation of AS gel at very early ages. In addition, the influence of Si modulus (Ms) of the activators and its relation with Na_2SiO_3:NaOH ratio was studied for maximizing the strength and to examine the economy of alkaline solution–activated binders in the preparation method [116,117]. The value of Ms determined the quantity of soluble silicates and controlled the dissolution rate and the gelation process during geopolymerization. This in turn significantly influenced the strength development of the hardened GP mixes. However, the value of Ms was varied for different GP systems obtained using various resource materials. Thus, varying chemical compositions were tried to examine the suitability of Ms for every group of GPs.

The influence of the Ms and varying concentration of alkali activator on the compressive strength of FA-based GP were evaluated by Guo et al. [118]. Mixtures of Na_2SiO_3 and NaOH were used as an activator where the silica–alkali modulus of the activator was changed from 1.0 up to 2.0. The concentration of alkali activator was dependent on the Na_2O to FA mass ratio between 5% and 15%. Results revealed ns that both silica–alkali modulus and alkali activator concentration were vital for the strength development of FA-based GPs. The optimum silica–alkali modulus and alkali activator contents were ascertained to be 1.5% and 10%, respectively. The compressive strength at 3, 7 and 28 days was 22.6, 34.5 and 59.3 MPa, respectively, for room temperature (23°C) curing. Law et al. [117] reported that the optimum Ms for Class F FA-based GP concrete is 1.0, where further increase in Ms did not show any considerable enlargement in the compressive strength. It was suggested that at Ms > 1.0, either all the FA particles were dissolved or any increase in Ms above 1.0 did not cause any further dissolution of the protective crust on the FA particles which was produced as precipitates from the geopolymerization reaction. Yusuf et al. [116] determined the compressive strength of alkali-activated GBFS-POFA–based GP under varying Ms in the range of 0.915–1.635. The value of Ms of 0.915 and 1.635 achieved the maximum compressive strength of 69.13 and 65 MPa, respectively.

Generally, compressive strength is related to the modulus of elasticity where higher rate of geopolymerization produces denser GP matrix, which in turn results in enhanced compressive strength and elastic modulus [119]. It is established that the chemical activator significantly influences the compressive strength development of GP concrete. However, other mechanical properties including elastic modulus of the GP concrete were somewhat independent on the activator amount [120]. In fact, the elastic modulus of the GP critically depends on the quantity of aggregates present in the GP mixes. An appropriate change in the total aggregate amount and the ratio of fine aggregate to total aggregate can give equal or higher elastic modulus of GP concrete as that of OPC concrete [121]. It was concluded that at very high Si concentration the elastic modulus of GP concrete may be lower than OPC concrete [42,122]. Topark-Ngarm et al. [119] demonstrated that high-calcium FA-based GP concrete has similar or higher elastic modulus with a reduced Na_2SiO_3 to NaOH ratio for higher amount of Na_2O.

3.6.2 Effect of Solution on Durability of Geopolymer

Ridtirud et al. [123] examined the influence of alkaline solution properties on shrinkage of FA GPs. The contraction of GPs is mainly influenced by the temperature of curing and liquid to FA proportion. At higher curing temperature, stronger GP with less shrinkage was obtained. Moreover, the shrinkage was increased considerably with an increase in liquid to FA ratios in the range of 0.4–0.7. Generally, this enhancement in the shrinkage was related to the low strength development of GPs. The contents of NaOH and Na_2SiO_3 to NaOH ratio have also influenced the shrinkage of GPs [6]. The influence of NaOH contents on the strength was weak but it was considerable on the shrinkage of GPs. NaOH contents of 12.5 M produced high shrinkage in GPs than the one with low NaOH contents of 7.5 M. The GP with high Na_2SiO_3 to NaOH ratio (3.0) revealed lower drying shrinkage than the one with lower ratios between 0.3 and 1.5. At higher Si concentration, the rate of reaction (condensation) was quite faster and the shrinkage was comparatively lower. Several researchers [124–126] have reported the increase in NaOH concentration and solution modulus enhance the durability of GP mortar by enhancing geopolymerization system. Enhanced geopolymerization improves the microstructure of GPMs and reduces the

TABLE 3.1
Properties of Different GPMs

Materials	High-Content Calcium-Based GPM Materials	High-Content Silicate-Based GPM Materials	High-Content Aluminium-Based GPM Materials
Reference	[5,9]	[9,127,128]	[129–133]
Binder	GBFS, FA-C	FA-F, POFA	MK
Setting time	Very short	Very long	Medium
Workability	Stiff	Very high	High
Early strength	Very high	Low	Medium
Durability	Medium	Very high	High
Curing temperature	Ambient	Oven	Oven
Suitability of repair	Very high	Low	Medium
Product cost	Low	High	Medium
Desirably	Very high	Low	High

FA, fly ash; MK, metakaolin.

water absorption, and increases the resistance to aggressive environments, such as acid and sulphate attack.

3.7 CHARACTERISTICS OF VARIOUS GEOPOLYMERS

Table 3.1 summarizes the characteristics of various GPMs. High-content calcium-based GPMs are more desirable as repair materials. They display very high performance with good properties. Furthermore, materials with high content of silicate as FA or POFA revealed improved workability and durability of GP.

3.8 GEOPOLYMER MIX DESIGN

Nowadays, GP binder–based industrial and agricultural wastes have been introduced as the environmentally friendly materials with high-durability performance. It is worth noting that millions of tons of natural, industrial and agriculture wastes such as FA, coal and oil-burning by-products, bottom ash, POFA, BA, used tyres, dust from cement, marble and crushed stone, waste ceramic materials, are dumped every year in landfills. These waste materials cause severe ecological setbacks such as air contamination and leach out of hazardous substances. Using these wastes in GP is a key part of decreasing present-day waste. GP industry saves natural resources and has a positive influence on cost-saving and environmental protection. Unlike traditional cement, there are many factors effect on GP mix design and future performance. The behaviour of proposed GP influenced by preparation method, binder chemical composition, molarity and modulus of alkaline activator solution. Finally, the ratio of binder to aggregates, alkaline solution to binder and curing method are considered the main factors effect on alkali-activated mix design. Table 3.2 presents different mixes of GPs as concrete repair materials. Clearly, it can be seen that there

TABLE 3.2
Mixture Designs of Geopolymer

Ref.	Type of Binder B1	B2	B3	SiO$_2$ to Al$_2$O$_3$	CaO to SiO$_2$	B:A*	Alkali Solution Type	NS:NH*	Molarity	S:B*
[5]	HFA	OPC	-							
	100	0	-	2.26	0.88	-	NHNS	2.0	6, 10, 14	1.0
	95	5	-	2.30	0.96	-				
	90	10	-	2.34	1.04	-				
	85	15	-	2.39	1.13	-				
[9]	FA	GBFS								
	100	0	-	1.93	0.06	-	NH	2.0	10	0.60
	50	50	-	2.03	0.60	-	NH			
	0	100	-	2.23	1.51	-	NH			
	100	0	-	1.93	0.06	-	NHNS			
	50	50	-	2.03	0.60	-	NHNS			
	0	100	-	2.23	1.51	-	NHNS			
	100	0	-	1.93	0.06	-	NS			
	50	50	-	2.03	0.60	-	NS			
	0	100	-	2.23	1.51	-	NS			
[133]	MK	FA	-							
	100	0		1.15	0.01	-	KHKS	-	-	-
	95	5		1.18	0.01					
	90	10		1.20	0.01					
[129]	MK	PAW	OPA							
	100	0	-	1.22	0.01	0.3	NHNS	2.5	-	0.83
	90	10	-	1.23	0.09					
	80	20	-	1.24	0.21					
	70	30	-	1.25	0.35					
	100	-	0	1.22	0.01					
	95	-	5	1.27	0.02					
	90	-	10	1.32	0.03					
	85	-	15	1.36	0.04					
[132]	MK	CaOH	-	-	-					-
	100	0	-	-	-	3,1.5,1	NHNS	2.5	12, 14, 16	
	95	5	-	-	-					
	90	10	-	-	-					
[127]	FA	GBFS	MSW							
	70	30		4.31	0.47	-	NHNS	-	-	
[130]	MK	GBFS								-
	100	0		1.71	0.01	-	NHNS	-	-	
	80	20		1.74	0.19	-	NHNS	-	-	
[128]	TMW	CaOH								-
	100	0		3.21	0	-	NHNS	2.5	24	
	90	10		3.31	0.19	-	NHNS	2.5	24	

B:A, binder to aggregate ratio; *NS:NH, sodium silicate to sodium hydroxide ratio; S:B, solution to binder ratio.

are many factors which play the main roles in GP design such as the binder, solution, filler and others.

3.9 SUMMARY

The following conclusions could be drawn based on the previous studies and results discussed in this chapter:

i. Several types of agriculture and by-product waste materials were re-used as binders to produce high-performance GP.
ii. The flowability, strength and durability of proposed GP were highly influenced by chemical composition and physical characterization of waste materials.
iii. In the design of GP mix, the content of alkaline solution, solution modulus and molarity depended on the type of binder and the content of calcium, silica, iron and alumina oxides.
iv. Materials containing low amount of CaO, such as FA and MK, need special curing regime and alkaline activator solution which highly effect on the sustainability of proposed GP.

REFERENCES

1. Ismail, M. and B. Muhammad, Electrochemical chloride extraction effect on blended cements. *Advances in Cement Research*, 2011. **23**(5): p. 241.
2. Mehmannavaz, T., et al., Binary effect of fly ash and palm oil fuel ash on heat of hydration aerated concrete. *The Scientific World Journal*, 2014. **2014**.
3. Bamaga, S., et al., Evaluation of sulfate resistance of mortar containing palm oil fuel ash from different sources. *Arabian Journal for Science and Engineering*, 2013. **38**(9): pp. 2293–2301.
4. Huseien, G.F., et al., Compressive strength and microstructure of assorted wastes incorporated geopolymer mortars: effect of solution molarity. *Alexandria Engineering Journal*, 2018. **57**(4): pp. 3375–3386.
5. Phoo-ngernkham, T., et al., High calcium fly ash geopolymer mortar containing Portland cement for use as repair material. *Construction and Building Materials*, 2015. **98**: pp. 482–488.
6. Huseien, G.F., et al., Geopolymer mortars as sustainable repair material: a comprehensive review. *Renewable and Sustainable Energy Reviews*, 2017. **80**: pp. 54–74.
7. Al-Majidi, M.H., et al., Development of geopolymer mortar under ambient temperature for in situ applications. *Construction and Building Materials*, 2016. **120**: pp. 198–211.
8. Huseien, G.F., et al., Influence of different curing temperatures and alkali activators on properties of GBFS geopolymer mortars containing fly ash and palm-oil fuel ash. *Construction and Building Materials*, 2016. **125**: pp. 1229–1240.
9. Phoo-ngernkham, T., et al., Effects of sodium hydroxide and sodium silicate solutions on compressive and shear bond strengths of FA–GBFS geopolymer. *Construction and Building Materials*, 2015. **91**: pp. 1–8.
10. Karakoç, M.B., et al., Mechanical properties and setting time of ferrochrome slag based geopolymer paste and mortar. *Construction and Building Materials*, 2014. **72**: pp. 283–292.

11. Pacheco-Torgal, F., J. Castro-Gomes, and S. Jalali, Alkali-activated binders: a review. Part 2. About materials and binders manufacture. *Construction and Building Materials*, 2008. **22**(7): pp. 1315–1322.
12. Horpibulsuk, S., R. Rachan, and Y. Raksachon, Role of fly ash on strength and microstructure development in blended cement stabilized silty clay. *Soils and Foundations*, 2009. **49**(1): pp. 85–98.
13. Huseien, G. F., M. A. Asaad, A. A. Abadel, S. K. Ghoshal, H. K. Hamzah, O. Benjeddou, and J. Mirza, Drying shrinkage, sulphuric acid and sulphate resistance of high-volume palm oil fuel ash-included alkali-activated mortars. *Sustainability*, 2022. **14**(1): p. 498.
14. Panias, D., I.P. Giannopoulou, and T. Perraki, Effect of synthesis parameters on the mechanical properties of fly ash-based geopolymers. *Colloids and Surfaces A: Physicochemical and Engineering Aspects*, 2007. **301**(1): pp. 246–254.
15. Cheng, T. and J. Chiu, Fire-resistant geopolymer produced by granulated blast furnace slag. *Minerals Engineering*, 2003. **16**(3): pp. 205–210.
16. Deb, P.S., P. Nath, and P.K. Sarker, The effects of ground granulated blast-furnace slag blending with fly ash and activator content on the workability and strength properties of geopolymer concrete cured at ambient temperature. *Materials & Design*, 2014. **62**: pp. 32–39.
17. Ismail, M., et al. Early strength characteristics of palm oil fuel ash and metakaolin blended geopolymer mortar. in *Advanced materials research*. 2013: Trans Tech Publ.
18. Khankhaje, E., et al., On blended cement and geopolymer concretes containing palm oil fuel ash. *Materials & Design*, 2016. **89**: pp. 385–398.
19. Salami, B.A., et al., Impact of added water and superplasticizer on early compressive strength of selected mixtures of palm oil fuel ash-based engineered geopolymer composites. *Construction and Building Materials*, 2016. **109**: pp. 198–206.
20. Salih, M.A., et al., Effect of different curing temperatures on alkali activated palm oil fuel ash paste. *Construction and Building Materials*, 2015. **94**: pp. 116–125.
21. Roy, D. Hydration, structure, and properties of blast furnace slag cements, mortars, and concrete. *Journal Proceedings*. 1982. **79**(6): pp. 444–457.
22. Turner, L.K. and F.G. Collins, Carbon dioxide equivalent (CO_2-e) emissions: a comparison between geopolymer and OPC cement concrete. *Construction and Building Materials*, 2013. **43**: pp. 125–130.
23. Provis, J.L., Geopolymers and other alkali activated materials: why, how, and what? *Materials and Structures*, 2014. **47**(1): pp. 11–25.
24. Weng, T.-L., W.-T. Lin, and A. Cheng, Effect of metakaolin on strength and efflorescence quantity of cement-based composites. *The Scientific World Journal*, 2013. **2013**: pp. 1–12.
25. Hawa, A., D. Tonnayopas, and W. Prachasaree, Performance evaluation and microstructure characterization of metakaolin-based geopolymer containing oil palm ash. *The Scientific World Journal*, 2013. **2013**: pp. 1–10.
26. Alengaram, U.J., B.A. Al Muhit, and M.Z. bin Jumaat, Utilization of oil palm kernel shell as lightweight aggregate in concrete–A review. *Construction and Building Materials*, 2013. **38**: pp. 161–172.
27. Antiohos, S. and S. Tsimas, A novel way to upgrade the coarse part of a high calcium fly ash for reuse into cement systems. *Waste Management*, 2007. **27**(5): pp. 675–683.
28. Bankowski, P., L. Zou, and R. Hodges, Reduction of metal leaching in brown coal fly ash using geopolymers. *Journal of Hazardous Materials*, 2004. **114**(1): pp. 59–67.
29. Yang, K.-H., et al., Properties and sustainability of alkali-activated slag foamed concrete. *Journal of Cleaner Production*, 2014. **68**: pp. 226–233.
30. Goodwin, R.W., *Combustion ash residue management: an engineering perspective*. 2013: William Andrew.
31. Fernández-Jiménez, A., A. Palomo, and M. Criado, Microstructure development of alkali-activated fly ash cement: a descriptive model. *Cement and Concrete Research*, 2005. **35**(6): pp. 1204–1209.

32. Bakharev, T., Geopolymeric materials prepared using class F fly ash and elevated temperature curing. *Cement and Concrete Research*, 2005. **35**(6): pp. 1224–1232.
33. Bakharev, T., Thermal behaviour of geopolymers prepared using class F fly ash and elevated temperature curing. *Cement and Concrete Research*, 2006. **36**(6): pp. 1134–1147.
34. Okoye, F., J. Durgaprasad, and N. Singh, Effect of silica fume on the mechanical properties of fly ash based-geopolymer concrete. *Ceramics International*, 2016. **42**(2): pp. 3000–3006.
35. Duan, P., et al., Fresh properties, mechanical strength and microstructure of fly ash geopolymer paste reinforced with sawdust. *Construction and Building Materials*, 2016. **111**: pp. 600–610.
36. Chindaprasirt, P., et al., High-strength geopolymer using fine high-calcium fly ash. *Journal of Materials in Civil Engineering*, 2010. **23**(3): pp. 264–270.
37. Hardjito, D., C.C. Cheak, and C.H.L. Ing, Strength and setting times of low calcium fly ash-based geopolymer mortar. *Modern Applied Science*, 2008. **2**(4): p. 3.
38. Sumajouw, D., et al., Fly ash-based geopolymer concrete: study of slender reinforced columns. *Journal of Materials Science*, 2007. **42**(9): pp. 3124–3130.
39. Bagheri, A. and A. Nazari, Compressive strength of high strength class C fly ash-based geopolymers with reactive granulated blast furnace slag aggregates designed by Taguchi method. *Materials & Design*, 2014. **54**: pp. 483–490.
40. Feng, J., et al., Development of porous fly ash-based geopolymer with low thermal conductivity. *Materials & Design*, 2015. **65**: pp. 529–533.
41. Rashad, A.M., Properties of alkali-activated fly ash concrete blended with slag. *Iranian Journal of Materials Science and Engineering*, 2013. **10**(1): pp. 57–64.
42. Sofi, M., et al., Engineering properties of inorganic polymer concretes (IPCs). *Cement and Concrete Research*, 2007. **37**(2): pp. 251–257.
43. Yusuf, M.O., et al., Evolution of alkaline activated ground blast furnace slag–ultrafine palm oil fuel ash based concrete. *Materials & Design*, 2014. **55**: pp. 387–393.
44. Yusuf, M.O., et al., Effects of H_2O/Na_2O molar ratio on the strength of alkaline activated ground blast furnace slag-ultrafine palm oil fuel ash based concrete. *Materials & Design*, 2014. **56**: pp. 158–164.
45. Nikolić, V., et al., The influence of fly ash characteristics and reaction conditions on strength and structure of geopolymers. *Construction and Building Materials*, 2015. **94**: pp. 361–370.
46. Lloyd, N. and B. Rangan. Geopolymer concrete with fly ash. in *Second international conference on sustainable construction materials and technologies*. 2010: UWM Centre for By-products Utilization.
47. Sarker, P. and V. Rangan, Geopolymer concrete using fly ash, in: *Applications and potential environmental impacts*. Edited by P. Sarker, Nova Science Publications, New York, 2014. pp. 271–289.
48. Zhang, Z., X. Yao, and H. Zhu, Potential application of geopolymers as protection coatings for marine concrete: II. Microstructure and anticorrosion mechanism. *Applied Clay Science*, 2010. **49**(1): pp. 7–12.
49. Temuujin, J., et al., Fly ash based geopolymer thin coatings on metal substrates and its thermal evaluation. *Journal of Hazardous Materials*, 2010. **180**(1): pp. 748–752.
50. Temuujin, J., et al., Preparation and thermal properties of fire resistant metakaolin-based geopolymer-type coatings. *Journal of Non-Crystalline Solids*, 2011. **357**(5): pp. 1399–1404.
51. Chindaprasirt, P., U. Rattanasak, and S. Taebuanhuad, Resistance to acid and sulfate solutions of microwave-assisted high calcium fly ash geopolymer. *Materials and Structures*, 2013. **46**(3): pp. 375–381.
52. Ariffin, M., et al., Sulfuric acid resistance of blended ash geopolymer concrete. *Construction and Building Materials*, 2013. **43**: pp. 80–86.

53. Tangchirapat, W., et al., Use of waste ash from palm oil industry in concrete. *Waste Management*, 2007. **27**(1): pp. 81–88.
54. Ranjbar, N., et al., Compressive strength and microstructural analysis of fly ash/palm oil fuel ash based geopolymer mortar. *Materials & Design*, 2014. **59**: pp. 532–539.
55. Ranjbar, N., et al., Compressive strength and microstructural analysis of fly ash/palm oil fuel ash based geopolymer mortar under elevated temperatures. *Construction and Building Materials*, 2014. **65**: pp. 114–121.
56. Awal, A.A. and M.W. Hussin, The effectiveness of palm oil fuel ash in preventing expansion due to alkali-silica reaction. *Cement and Concrete Composites*, 1997. **19**(4): pp. 367–372.
57. Alengaram, U.J., H. Mahmud, and M.Z. Jumaat, Enhancement and prediction of modulus of elasticity of palm kernel shell concrete. *Materials & Design*, 2011. **32**(4): pp. 2143–2148.
58. Mo, K.H., U.J. Alengaram, and M.Z. Jumaat, A review on the use of agriculture waste material as lightweight aggregate for reinforced concrete structural members. *Advances in Materials Science and Engineering*, 2014. **2014**.
59. Awal, A.A. and M.W. Hussin, Effect of palm oil fuel ash in controlling heat of hydration of concrete. *Procedia Engineering*, 2011. **14**: pp. 2650–2657.
60. Chindaprasirt, P., S. Rukzon, and V. Sirivivatnanon, Resistance to chloride penetration of blended Portland cement mortar containing palm oil fuel ash, rice husk ash and fly ash. *Construction and Building Materials*, 2008. **22**(5): pp. 932–938.
61. Salami, B.A., et al., Durability performance of Palm Oil Fuel Ash-based Engineered Alkaline-activated Cementitious Composite (POFA-EACC) mortar in sulfate environment. *Construction and Building Materials*, 2017. **131**: pp. 229–244.
62. Salih, M.A., A.A.A. Ali, and N. Farzadnia, Characterization of mechanical and microstructural properties of palm oil fuel ash geopolymer cement paste. *Construction and Building Materials*, 2014. **65**: pp. 592–603.
63. Islam, A., et al., The development of compressive strength of ground granulated blast furnace slag-palm oil fuel ash-fly ash based geopolymer mortar. *Materials & Design*, 2014. **56**: pp. 833–841.
64. Karim, M., et al., Fabrication of a non-cement binder using slag, palm oil fuel ash and rice husk ash with sodium hydroxide. *Construction and Building Materials*, 2013. **49**: pp. 894–902.
65. Salih, M.A., et al., Development of high strength alkali activated binder using palm oil fuel ash and GGBS at ambient temperature. *Construction and Building Materials*, 2015. **93**: pp. 289–300.
66. Yusuf, T.O., et al., Impact of blending on strength distribution of ambient cured metakaolin and palm oil fuel ash based geopolymer mortar. *Advances in Civil Engineering*, 2014. **2014**: pp. 1–9.
67. Bakharev, T., Durability of geopolymer materials in sodium and magnesium sulfate solutions. *Cement and Concrete Research*, 2005. **35**(6): pp. 1233–1246.
68. Temuujin, J., A. van Riessen, and K. MacKenzie, Preparation and characterisation of fly ash based geopolymer mortars. *Construction and Building Materials*, 2010. **24**(10): pp. 1906–1910.
69. Hussin, M., et al., Performance of blended ash geopolymer concrete at elevated temperatures. *Materials and Structures*, 2015. **48**(3): pp. 709–720.
70. Izquierdo, M., et al., Coal fly ash-slag-based geopolymers: microstructure and metal leaching. *Journal of Hazardous Materials*, 2009. **166**(1): pp. 561–566.
71. Nath, P. and P.K. Sarker, Effect of GGBFS on setting, workability and early strength properties of fly ash geopolymer concrete cured in ambient condition. *Construction and Building Materials*, 2014. **66**: pp. 163–171.
72. Sahana, R. Setting time compressive strength and microstructure of geopolymer paste. in *Proceedings of the International Conference on Energy and Environment (ICEE'13)*. 2013.

73. Marjanović, N., et al., Physical–mechanical and microstructural properties of alkali-activated fly ash–blast furnace slag blends. *Ceramics International*, 2015. **41**(1): pp. 1421–1435.
74. Nath, P., P.K. Sarker, and V.B. Rangan, Early age properties of low-calcium fly ash geopolymer concrete suitable for ambient curing. *Procedia Engineering*, 2015. **125**: pp. 601–607.
75. Puertas, F., et al., Alkali-activated fly ash/slag cements: strength behaviour and hydration products. *Cement and Concrete Research*, 2000. **30**(10): pp. 1625–1632.
76. Shi, C. and R. Day, Early strength development and hydration of alkali-activated blast furnace slag/fly ash blends. *Advances in Cement Research*, 1999. **11**(4): pp. 189–196.
77. Kürklü, G., The effect of high temperature on the design of blast furnace slag and coarse fly ash-based geopolymer mortar. *Composites Part B: Engineering*, 2016. **92**: pp. 9–18.
78. Chindaprasirt, P., et al., Effect of SiO_2 and Al_2O_3 on the setting and hardening of high calcium fly ash-based geopolymer systems. *Journal of Materials Science*, 2012. **47**(12): pp. 4876–4883.
79. Raijiwala, D. and H. Patil, Geopolymer concrete: a concrete of the next decade. *Concrete Solutions* 2011, 2011: p. 287.
80. Lee, N., E. Kim, and H. Lee, Mechanical properties and setting characteristics of geopolymer mortar using styrene-butadiene (SB) latex. *Construction and Building Materials*, 2016. **113**: pp. 264–272.
81. Li, Z. and S. Liu, Influence of slag as additive on compressive strength of fly ash-based geopolymer. *Journal of Materials in Civil Engineering*, 2007. **19**(6): pp. 470–474.
82. Guerrieri, M. and J.G. Sanjayan, Behavior of combined fly ash/slag-based geopolymers when exposed to high temperatures. *Fire and Materials*, 2010. **34**(4): pp. 163–175.
83. Ismail, I., et al., Microstructural changes in alkali activated fly ash/slag geopolymers with sulfate exposure. *Materials and Structures*, 2013. **46**(3): pp. 361–373.
84. Chi, M. and R. Huang, Binding mechanism and properties of alkali-activated fly ash/slag mortars. *Construction and Building Materials*, 2013. **40**: pp. 291–298.
85. Fernandes, M., A. Sousa, and A. Dias, *Environmental impacts and emissions trading-ceramic industry: a case study*. 2004: Technological centre of ceramics and glass, Portuguese association of ceramic industry (in Portuguese).
86. Pacheco-Torgal, F. and S. Jalali, Reusing ceramic wastes in concrete. *Construction and Building Materials*, 2010. **24**(5): pp. 832–838.
87. Limbachiya, M., M.S. Meddah, and Y. Ouchagour, Use of recycled concrete aggregate in fly-ash concrete. *Construction and Building Materials*, 2012. **27**(1): pp. 439–449.
88. Heidari, A. and D. Tavakoli, A study of the mechanical properties of ground ceramic powder concrete incorporating nano-SiO_2 particles. *Construction and Building Materials*, 2013. **38**: pp. 255–264.
89. Samadi, M., et al., Properties of mortar containing ceramic powder waste as cement replacement. *Jurnal Teknologi*, 2015. **77**(12): pp. 93–97.
90. Senthamarai, R., P.D. Manoharan, and D. Gobinath, Concrete made from ceramic industry waste: durability properties. *Construction and Building Materials*, 2011. **25**(5): pp. 2413–2419.
91. Ariffin, M.A., et al., Effect of ceramic aggregate on high strength multi blended ash geopolymer mortar. *Jurnal Teknologi*, 2015. **77**(16): pp. 33–36.
92. Brough, A. and A. Atkinson, Sodium silicate-based, alkali-activated slag mortars: part I. Strength, hydration and microstructure. *Cement and Concrete Research*, 2002. **32**(6): pp. 865–879.
93. Granizo, M.L., et al., Alkaline activation of metakaolin: effect of calcium hydroxide in the products of reaction. *Journal of the American Ceramic Society*, 2002. **85**(1): pp. 225–231.

94. Deir, E., B.S. Gebregziabiher, and S. Peethamparan, Influence of starting material on the early age hydration kinetics, microstructure and composition of binding gel in alkali activated binder systems. *Cement and Concrete Composites*, 2014. **48**: pp. 108–117.
95. Rashad, A.M., A comprehensive overview about the influence of different admixtures and additives on the properties of alkali-activated fly ash. *Materials & Design*, 2014. **53**: pp. 1005–1025.
96. Duan, P., et al., An investigation of the microstructure and durability of a fluidized bed fly ash–metakaolin geopolymer after heat and acid exposure. *Materials & Design*, 2015. **74**: pp. 125–137.
97. Makhloufi, Z., et al., Effect of quaternary cementitious systems containing limestone, blast furnace slag and natural pozzolan on mechanical behavior of limestone mortars. *Construction and Building Materials*, 2015. **95**: pp. 647–657.
98. Yusuf, M.O., et al., Influence of curing methods and concentration of NaOH on strength of the synthesized alkaline activated ground slag-ultrafine palm oil fuel ash mortar/concrete. *Construction and Building Materials*, 2014. **66**: pp. 541–548.
99. Puertas, F., A. Fernández-Jiménez, and M.T. Blanco-Varela, Pore solution in alkali-activated slag cement pastes. Relation to the composition and structure of calcium silicate hydrate. *Cement and Concrete Research*, 2004. **34**(1): pp. 139–148.
100. Palacios, M. and F. Puertas, Effect of shrinkage-reducing admixtures on the properties of alkali-activated slag mortars and pastes. *Cement and Concrete Research*, 2007. **37**(5): pp. 691–702.
101. Gao, X., Q. Yu, and H. Brouwers, Reaction kinetics, gel character and strength of ambient temperature cured alkali activated slag–fly ash blends. *Construction and Building Materials*, 2015. **80**: pp. 105–115.
102. Gao, X., Q. Yu, and H. Brouwers, Properties of alkali activated slag–fly ash blends with limestone addition. *Cement and Concrete Composites*, 2015. **59**: pp. 119–128.
103. Gao, X., Q. Yu, and H. Brouwers, Characterization of alkali activated slag–fly ash blends containing nano-silica. *Construction and Building Materials*, 2015. **98**: pp. 397–406.
104. Papa, E., et al., Production and characterization of geopolymers based on mixed compositions of metakaolin and coal ashes. *Materials & Design*, 2014. **56**: pp. 409–415.
105. Hu, M., X. Zhu, and F. Long, Alkali-activated fly ash-based geopolymers with zeolite or bentonite as additives. *Cement and Concrete Composites*, 2009. **31**(10): pp. 762–768.
106. Part, W.K., M. Ramli, and C.B. Cheah, An overview on the influence of various factors on the properties of geopolymer concrete derived from industrial by-products. *Construction and Building Materials*, 2015. **77**: pp. 370–395.
107. Komljenović, M., Z. Baščarević, and V. Bradić, Mechanical and microstructural properties of alkali-activated fly ash geopolymers. *Journal of Hazardous Materials*, 2010. **181**(1): pp. 35–42.
108. Lin, K.L., et al., Effects of SiO_2/Na_2O molar ratio on properties of TFT-LCD waste glass-metakaolin-based geopolymers. *Environmental Progress & Sustainable Energy*, 2014. **33**(1): pp. 205–212.
109. De Silva, P., K. Sagoe-Crenstil, and V. Sirivivatnanon, Kinetics of geopolymerization: role of Al_2O_3 and SiO_2. *Cement and Concrete Research*, 2007. **37**(4): pp. 512–518.
110. North, M.R. and T.W. Swaddle, Kinetics of silicate exchange in alkaline aluminosilicate solutions. *Inorganic Chemistry*, 2000. **39**(12): pp. 2661–2665.
111. Ghosh, K. and P. Ghosh, Effect Of Na_2O/Al_2O_3, SiO_2/Al_2O_3 and w/b ratio on setting time and workability of fly ash based geopolymer. *International Journal of Engineering Research and Applications*, 2012. **2**(4): pp. 2142–2147.
112. Sathonsaowaphak, A., P. Chindaprasirt, and K. Pimraksa, Workability and strength of lignite bottom ash geopolymer mortar. *Journal of Hazardous Materials*, 2009. **168**(1): pp. 44–50.
113. Sukmak, P., S. Horpibulsuk, and S.-L. Shen, Strength development in clay–fly ash geopolymer. *Construction and Building Materials*, 2013. **40**: pp. 566–574.

114. Görhan, G. and G. Kürklü, The influence of the NaOH solution on the properties of the fly ash-based geopolymer mortar cured at different temperatures. *Composites Part B: Engineering*, 2014. **58**: pp. 371–377.
115. Somna, K., et al., NaOH-activated ground fly ash geopolymer cured at ambient temperature. *Fuel*, 2011. **90**(6): pp. 2118–2124.
116. Yusuf, M.O., et al., Impacts of silica modulus on the early strength of alkaline activated ground slag/ultrafine palm oil fuel ash based concrete. *Materials and Structures*, 2015. **48**(3): pp. 733–741.
117. Law, D.W., et al., Long term durability properties of class F fly ash geopolymer concrete. *Materials and Structures*, 2015. **48**(3): pp. 721–731.
118. Guo, X., H. Shi, and W.A. Dick, Compressive strength and microstructural characteristics of class C fly ash geopolymer. *Cement and Concrete Composites*, 2010. **32**(2): pp. 142–147.
119. Topark-Ngarm, P., P. Chindaprasirt, and V. Sata, Setting time, strength, and bond of high-calcium fly ash geopolymer concrete. *Journal of Materials in Civil Engineering*, 2014. **27**(7): p. 04014198.
120. Khandelwal, M., et al., Effect of strain rate on strength properties of low-calcium fly-ash-based geopolymer mortar under dry condition. *Arabian Journal of Geosciences*, 2013. **6**(7): pp. 2383–2389.
121. Joseph, B. and G. Mathew, Influence of aggregate content on the behavior of fly ash based geopolymer concrete. *Scientia Iranica*, 2012. **19**(5): pp. 1188–1194.
122. Olivia, M. and H. Nikraz, Properties of fly ash geopolymer concrete designed by Taguchi method. *Materials & Design*, 2012. **36**: pp. 191–198.
123. Ridtirud, C., P. Chindaprasirt, and K. Pimraksa, Factors affecting the shrinkage of fly ash geopolymers. *International Journal of Minerals, Metallurgy, and Materials*, 2011. **18**(1): pp. 100–104.
124. Ahmari, S. and L. Zhang, Production of eco-friendly bricks from copper mine tailings through geopolymerization. *Construction and Building Materials*, 2012. **29**: pp. 323–331.
125. Chindaprasirt, P., U. Rattanasak, and C. Jaturapitakkul, Utilization of fly ash blends from pulverized coal and fluidized bed combustions in geopolymeric materials. *Cement and Concrete Composites*, 2011. **33**(1): pp. 55–60.
126. Freidin, C., Cementless pressed blocks from waste products of coal-firing power station. *Construction and Building Materials*, 2007. **21**(1): pp. 12–18.
127. Dai, Y., et al., *A study on application of geopolymeric green cement*. National Ilan University, 2013: pp. 1–8.
128. Torgal, F.P., J. Gomes, and S. Jalali, Bond strength between concrete substance and repair materials: comparisons between tungsten mine waste geopolymeric binder versus current commercial repair products. in: *17th International Congress on Advances in Civil Engineering*. October 2006, Yildiz Technical University, Istanbul, Turkey: pp. 1–10.
129. Hawa, A., et al., Development and performance evaluation of very high early strength geopolymer for rapid road repair. *Advances in Materials Science and Engineering*, 2013. **2013**: pp. 1–10.
130. Hu, S., et al., Bonding and abrasion resistance of geopolymeric repair material made with steel slag. *Cement and Concrete Composites*, 2008. **30**(3): pp. 239–244.
131. Moura, D., et al. Concrete repair with geopolymeric mortars: influence of mortars composition on their workability and mechanical strength. in *VI International Materials Symposium (Materials 2011)*. 2011.
132. Vasconcelos, E., et al. Concrete retrofitting using CFRP and geopolymer mortars. in *Materials Science Forum*. 2013. Trans Tech Publ.
133. Zhang, H.Y., et al., Characterizing the bond strength of geopolymers at ambient and elevated temperatures. *Cement and Concrete Composites*, 2015. **58**: pp. 40–49.

4 Factors Effect on the Manufacturing of Geopolymer

4.1 INTRODUCTION

As mentioned in previous chapters, geopolymer (GP) comprises SiO_2 and Al_2O_3 resource components with concentrated alkali solution activation [1]. Several complementary cementitious systems are generally utilized as resource materials for GP including ground blast furnace slag (GBFS), metakaolin (MK) and fly ash (FA) due to their abundance and positive mechanical attributes [2–6]. It has been reported [7–10] that FA with high Ca contents is advantageous for preparing high-strength GPs. Although high calcium content FA-based geopolymer mortars (GPMs) can be cured at room temperature, extremely slow rate of geopolymerization at ambient condition without additives [10–12] is responsible for their low binding strength. In this regard, Ordinary Portland Cement (OPC) is superior to enhancing the strength of high calcium FA-based GP [8,13–15]. In addition to the formation of calcium–silicate–hydrate (C-S-H), the heat generation from ordinary Portland cement (OPC) and water-assisted geopolymerization process can enhance their bonding strength [16]. GPM with high calcium content was prepared by adding OPC at 25°C and a compressive strength of 65 MPa was achieved [8].

Tanakorn et al. [17] studied the effect of calcium content on the bonding strength of GP. They used FA (Class C) with high calcium content in various ratios of OPC in the range of 0%–15% [6]. GBFS and Class F FA (low calcium content) together with three types of geopolymer paste (GPP) such as FA, FA + GBFS and GBFS were examined. The studied ratio of FA to GBFS is 100%, 50% and 0%. Zhang et al. (2015) [18] reported the influence of high-content alumina materials on the bonding strength of MK. Commercial MK (Shanxi Jinkunhengye Co., Ltd., China) was calcined at 900°C to prepare the test material where GP samples made from pure MK were more susceptible to shrinkage [19–22]. This shrinkage is further minimized by partially substituting MK with FA (Grade I: Chinese Code GB/T 1596–2005 specifications) [18]. The chemical composite and mixing design of MK and FA are listed in Tables 2.3 and 2.4.

Hawa et al. [23] produced GPM from MK by mixing it with para-wood ash (PWA, rubber-wood ash) or oil palm ash (OPA) as binder agent. The chemical composition of three waste materials and their mix design is summarized in Table 4.1. Vasconcelos et al. [24] reported the use of MK as a binder without any waste materials. MK is mixed with different binder-to-aggregate ratios and molarity (Table 2.3). Moura et al. [25] studied the effect of calcium hydroxide replaced by MK-based GP at 5%

TABLE 4.1
Chemical Composition of Various GP Mortars Obtained from XRF Test

Ref.	Materials	SiO$_2$	Al$_2$O$_3$	Fe$_2$O$_3$	CaO	MgO	K$_2$O	Na$_2$O	SO$_3$	LOI
[6]	HFA	29.32	12.96	15.64	25.79	2.94	2.93	2.83	7.29	0.30
	OPC	20.80	4.70	3.40	65.30	1.50	0.40	0.10	2.70	0.90
[17]	FA	52.31	27.04	6.85	3.32	1.23	1.29	1.15	0.99	1.60
	GBFS	30.53	13.67	0.33	46.0	5.09	0.36	0.24	-	0.22
[18]	MK	51.85	44.84	0.98	0.13	0.48	0.08	0.16	-	-
	FA	48.66	24.80	21.10	3.30	1.10	-	-	1.04	-
[23]	MK	50.30	41.02	1.05	0.33	-	4.08	-	-	1.72
	PAW	2.57	0.53	0.56	41.19	4.52	16.11	-	5.54	23.74
	OPA	38.37	1.48	3.01	13.84	3.0	14.09	-	1.42	20.43
[24]	MK	-	-	-	-	-	-	-	-	-

XRF, X-ray fluorescence.

and 10% under different alkali solution concentrations (Table 4.1). Dai et al. [26] reported the use of GBFS and FA obtained from the CHC Resources Corporation. The waste in this study is a municipal solid waste incinerator (MSWI) FA from an incineration treatment plant in Northern Taiwan. Hu et al. [27] investigated the effect of ground-granulated blast furnace slag replaced by MK (0% and 20%) on the bonding and the abrasion resistance (AR) of GPM. Torgal et al. [28] studied the implementation of tungsten mine waste (TMW) geopolymeric binder (mine waste mud – TMW) and calcium hydroxide at 10% substitution.

For fine aggregate–based GP, the river sand is mostly used. The specific gravity of river sand is between 2.5 and 2.6 and the maximum size is 4.75 mm. According to ASTM C128 test, water absorption of river sand is about 1.41%–1.48%. To activate the GP binders, most of the reports acknowledged that alkaline activator is a mixture of sodium hydroxide (NaOH with 98% purity) and sodium silicate (Na$_2$SiO$_3$). Usually, the Na$_2$SiO$_3$ solution is characterized by SiO$_2$:Na$_2$O weight ratio varied within 2–3.75, where a value greater than 2.85 signifies a neutral solution [10]. Only few researchers used commercially available K$_2$SiO$_3$ solution with 15.8 wt% of K$_2$O, 24.2 wt% of SiO$_2$ and 60 wt% of H$_2$O (SiO$_2$:K$_2$O molar ratio was 2.4), KOH flakes (85% pure) and tap water to prepare GPs [18]. In this chapter, the influence of GP flowability, compressive, flexural, tensile and bond strength, drying shrinkage by the binder type, binder to solution and filler ratios, alkaline activator solution modulus and molarity were reviewed.

4.2 FRESH PROPERTIES OF GEOPOLYMER

Recently, the workability of GPM as repair materials (RMs) was examined. Tanakorn et al. [17] inspected the effect of sodium hydroxide and ratio of calcium oxide to silicate on the workability of GPM. In this study, the molarities are varied to 6, 10 and 14 M. To study the impact of calcium oxide-to-silicate ratio, FA Class C was replaced by OPC in the range of 0%–15% as summarized in Table 4.2. It is clear that an increase

TABLE 4.2
Effect Calcium Content and NaOH Concentration on Setting Time of GMP [17]

| | NaOH (6 M) | | NaOH (10 M) | | NaOH (14 M) | |
| | Setting Time (min) | | Setting Time (min) | | Setting Time (min) | |
POC (%)	i.s[a]	f.s[a]	i.s	f.s	i.s	f.s
0	21	50	47	88	80	130
5	14	25	30	45	65	105
10	12	19	15	30	23	40
15	7	12	10	20	18	25

a(i.s) initial setting time, (f.s) final setting time.

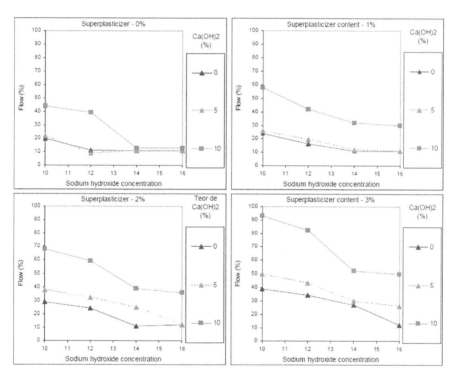

FIGURE 4.1 NaOH content–dependent flow of mortars under different super-plasticizer and $Ca(OH)_2$ concentration [25].

in sodium hydroxide molarity had increased the initial (i.s) and final (f.s) setting time. The effect of calcium content on setting time was also reported. It was found that an increased percentage of OPC replaced by FA led to a reduction of the initial and final setting time of GPM. The addition of steel slag could accelerate the setting time and significantly improve the compressive strength, which was attributed to its latent hydraulic cementitious character [27].

Moura et al. [25] studied the effects of varying super-plasticizer contents (1%, 2%, and 3%) on the GPM setting time. The activation of alkaline solution in the GPMs in the absence of super-plasticizer showed the flow below 50%. However, the GPMs with higher super-plasticizer contents displayed an enhanced flow. GPMs in the presence of high $Ca(OH)_2$ and low percentage of MK showed a high flow as illustrated in Figure 4.1. The high Blaine fineness of MK required a higher amount of fluid for better solubility. Flow is also reduced at high sodium hydroxide concentration. The highest flow is achieved by mortars at a sodium hydroxide concentration of 10 M and a calcium hydroxide of 10%. The use of a super-plasticizer content of 3% combined with a $Ca(OH)_2$ content of 10% enhanced mortar flow from less than 50% to over 90% [29].

4.3 COMPRESSIVE STRENGTH

The compressive strength of concrete is indeed the most valuable physical property, where high early strength is significant for RMs. Features such as tensile and flexural strength, and elastic modulus of GPMs were observed to depend on the compressive strength which was measured using ASTM C109/C109M. Factors that affect the GPM strength are binder-to-aggregate ratio, molarity of sodium hydroxide, sodium silicate-to-sodium hydroxide ratio, solution-to-binder ratio, silicate-to-Al ratio, silicate-to-sodium oxide and calcium content. Some of these factors are described below.

4.3.1 Effect of Calcium Content

Tanakon et al. [17] reported the influence of calcium ratio on the binding feature of GPM, where FA (high-content calcium) and OPC are used. The OPC is replaced by FA with 0%, 5%, 10% and 15%. An increase in calcium content from 6 to 10 M is found to increase the compressive strength (Table 4.3). Furthermore, an increase in calcium content to high alkali of 14 M enhanced the compressive strength up to 10% and reduced the strength thereafter. Phoongernkham et al. [6] studied the effect of high calcium content, where FA was replaced by GBFS at 0%, 50% and 100%.

TABLE 4.3
Properties and Strength of GPM [17]

	NaOH (6 M)				NaOH (10 M)				NaOH (14 M)			
	Setting Time (min)		GPM Strength (MPa)		Setting Time (min)		GPM Strength (MPa)		Setting Time (min)		GPM Strength (MPa)	
Symbols	i.s	f.s	f_c	f_t	i.s	f.s	f_c^a	f_t^a	i.s	f.s	f_c	f_t
OPC	21	50	38.5	2.91	47	88	50.5	6.22	80	130	56.0	7.07
5PC	14	25	40.2	3.79	30	45	56.7	6.85	65	105	58.5	8.51
10PC	12	19	45.3	4.77	15	30	60.4	7.17	23	40	63.3	8.96
15PC	7	12	48.2	4.93	10	20	64.1	7.32	18	25	62.0	7.49

[a](f_c) compressive strength, (f_t) flexural strength.

Figure 4.2 displays the binder-dependent compressive strengths of various GPs (GPPs). The compressive strength was found to increase with the increase in GBFS content for every alkaline solution. This enhancement can be ascribed to the promptly accessible free calcium ions that reacted with SiO_2 and Al_2O_3 to form C(A)-S-H gel and co-existed with GP gels [6]. Additionally, the exothermal reaction between GBFS and alkaline solutions liberated excess heat and promoted the rate of geopolymerization. Thus, an increasing GBFS concentration enhanced the compressive strength of GPs [6].

Hawa et al. [23] used PWA with high calcium content to determine its effects on the compressive strength of MK-based GPM which are depicted in Figure 4.3. The compressive strength at early 2 hours was observed to be high. This was partly because of the fact that the GPMs were prepared as a hot mixture before curing in an oven. Moreover, a decrease in the compressive strength with increasing PWA

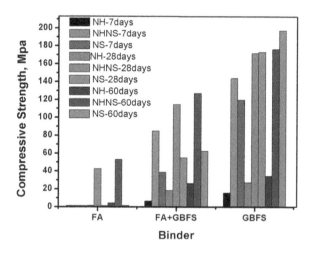

FIGURE 4.2 Binder-dependent compressive strength of GPs [6].

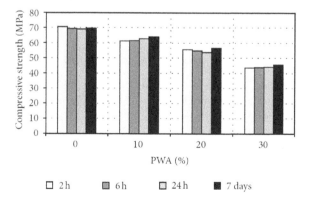

FIGURE 4.3 PWA-dependent compressive strength percentage for different mixtures with curing age of 4 hours in an oven [34].

TABLE 4.4
Changing Curing Age-Dependent Compressive Strength of Different RMs [27]

Mixture Types	Compressive Strength (MPa)				
	8 hours	1 day	3 days	7 days	28 days
Cement repair (CR)	-	8.7	23.2	33.1	46.1
Geopolymeric repair without steel slag (GR)	10.4	17.2	30.4	37.8	40.9
Geopolymeric repair with steel slag (GRS)	14.3	22.1	35.6	40.4	44.5

content was attributable to the particle size effects (6.31 µm of MK against 25.13 µm of PWA). The larger surface-to-volume ratio of finer particles was responsible for active polymerization [30,31]. The reduction of Si and Al when MK was replaced with PWA may be another contributing factor for the decrease. Actually, the CaO in PWA did not participate in geopolymerization and got hydrated slowly. Winnefeld et al. [32] found that lower strength incurred by high calcium content was due to poor reactivity with alkaline activators in FA-based GPs. It was confirmed that by adding CaO into raw materials, the compressive strength could be reduced. Promising results were found [33] at 70°C curing temperature.

Moura et al. [25] examined the effect of CaO replacement in MK-based GP with 0%, 5% and 10%. The replacement of MK by CaO was found to develop a maximum compressive strength for 10%. Dai et al. [26] studied the effect of high-content calcium on GPM, where GBFS and FA and MSWI FA were used as a waste binder. Hu et al. [27] reported the influence of GBFS replacement on MK-based GPM with 0% and 20%. The compressive strength at 8 hours, 1 day, 3, 7, and 28 days for different repairing systems are furnished in Table 4.4. The compressive strength of OPC as a RM was lower compared to GPMs both in the absence (geopolymeric repair (GR)) and presence of steel slag (geopolymeric repair with steel slag (GRS)). The achievement of elevated early compressive strength of GPMs was attributed to the strong alkaline activation. A comparison of GRS with GR revealed that at 8 hours, 1 day, 3, 7 and 28 days, the compressive strengths were increased by 43%, 28%, 17%, 6.9% and 7.6%, respectively.

4.3.2 Effect of Alkaline Solution Characterization

The sodium hydroxide molarity is one of the main factors effect on proposed GP engineering properties and sustainability performance. Tanakon et al. [17] reported the effects of varying NaOH molarities (6, 10 and 14 M) on developed compressive strength. The result indicated an increase in the compressive strength with an increase in molarity. Vasconcelos et al. [24] inspected the influence of changing NaOH concentration (12, 14 and 16 M) on strength development. Again, the compressive strength was enhanced with increasing NaOH content from 12 to 14 M, after 7, 28 and 56 days of curing. However, the strength was reduced after increasing the NaOH molarity above 14 M. Strength with molarity 16 M showed low strength compared with 14 M (Figure 4.4).

Factors Effect on Geopolymer

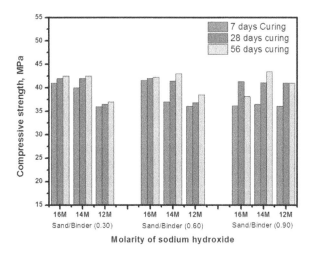

FIGURE 4.4 Effect of NaOH concentration and sand-to-binder ratio on the compressive strength of GPMs [24].

Phoongernkham et al. [6] studied the influence of solution types on the activation process of waste materials containing great amount of aluminium silicate. Three kinds of solutions such as NaOH (NH), Na_2SiO_3 (NS) and NaOH plus Na_2SiO_3 (NHNS) were considered. Figure 4.4 clearly displays that the GP mix with 100% of FA produced little early strength for all three series. The presence of NH solution played a critical role in dissolving the Si^{4+} and Al^{3+} ions and succeeding geopolymerization at ambient temperature [10]. Nonetheless, strength development of NH- or NS-activated FA paste was very weak [10,35]. However, NHNS-activated specimen displayed considerable development in the strength afterwards. The 7-day compressive strength of FA paste activated with NHNS was very low. Moreover, the compressive strengths were comparatively higher for samples cured at 28 days (45.0 MPa) and 60 days (52.9 MPa). The observed lesser strength of FA paste at early age was attributed to the slower reaction rate. Interestingly, the curing age-dependent strength development of the FA paste was comparable to that of the POC concrete [33].

NaOH-activated GP blend with higher GBFS level (FA + GBFS paste) produced poorer strength than the specimens activated with NHNS and NS solution. It is worth noting that NH was required to leach SiO_2 and Al_2O_3 from the system where the leaching and consequent reaction was slowed down at 23°C [36]. The use of NS only or combined with NH produced extra silicates and accelerated the geopolymerization process which finally led to the enhancement of the compressive strength. Besides, the NHNS-activated FA plus GBFS paste revealed superior strength at all ages than those activated with NH and NS. The NHNS alkaline solution–activated FA plus GBFS blend was considered to be the best binder with elevated strength.

High-strength GBFS–based GP mixes were also obtained with Na_2SO_4 [37]. The observed maximum compressive strength at 28 and 60 days age as revealed by GBFS paste activated with NS was ascribed to the reaction between CaO and SiO_2

and succeeding production of C-S-H gels [38]. Ismail et al. [38] studied that the alkaline solution–activated GBFS and showed the dissolution of Ca and contribution of Si and Al to generate C-S-H and C-A-S-H gel which produced large compressive strength. The compressive strength mainly depended on the molar ratio of $SiO_2:Al_2O_3$ [39]. An $SiO_2:Al_2O_3$ ratio of 3.50 produced a large strength for high Ca content GP [30]. The optimum value of $SiO_2:Al_2O_3$ ratio for GBFS paste activated with NS was ascertained to be 3.49 where the strength after 28 days of curing at ambient temperature was as much as 171.7 MPa. Regarding to ratio of sodium silicate to sodium hydroxide, most researchers have dealt with GPM as RMs, where the molar ratios of sodium silicate to sodium hydroxide are kept constant at 2.5 [23–25,28] and 2.0 [6,17]. However, sodium silicate-to-sodium hydroxide concentration ratio effect on the strength development of GP as RMs is far from being investigated.

4.3.3 Effect of Aggregate-to-Binder Ratio

Vasconcelos et al. [24] scrutinized the effect of aggregate-to-binder ratio (0.3, 0.60 and 0.9) on the strength improvement of GPM. Compressive strength showed an increase with increasing aggregate-to-binder ratio (Figure 4.4). The optimum result was achieved with 0.90 of sand-to-binder ratio and 14 M molarity of sodium hydroxide.

4.3.4 Effect $H_2O:Na_2O$ Ratio

Vasconcelos et al. [24] reported the effect of varying $H_2O:Na_2O$ ratio (9, 9.5 and 10) on the development of compressive strength of GPM studies to investigate the effect. Figure 4.5 illustrates that the compressive strength is decreased with the increase in the $H_2O:Na_2O$ ratio.

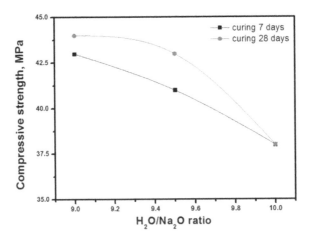

FIGURE 4.5 $H_2O:Na_2O$ atomic ratio–dependent compressive strength of GPMs with different sand-to-binder mass ratios of 30% and 60% according to curing days [24].

Factors Effect on Geopolymer

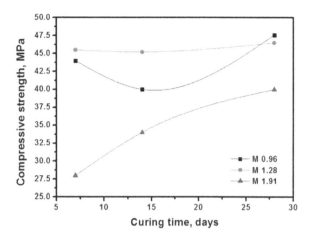

FIGURE 4.6 Curing time–dependent compressive strength of GPM at different $SiO_2:Na_2O$ ratios [26].

4.3.5 $SiO_2:Na_2O$ Ratio Effect

Dai et al. [26] inspected the effect of varying $SiO_2:Na_2O$ ratio (0.96, 1.28 and 1.91) on the development of GPM compressive strength (Figure 4.6). The mechanical strength was found to increase with increasing curing time which was about 47.1 MPa for $SiO_2:Na_2O$ of 0.96. Furthermore, the observed reduction in the compressive strength with increasing $SiO_2:Na_2O$ molar ratio was ascribed to participation of excessive Si ions in geopolymerization. The lowest compressive strength (40.8 MPa) was obtained at $SiO_2:Na_2O$ ratio of 1.91. Generally, the superior strength of geopolymeric green cement can be ascribed to the formation of alumina-silicate monopolymer during geopolymeric reaction at an early age. At lower $SiO_2:Na_2O$ molar ratio (0.96 and 1.28), the occurrence of higher compressive strength was due to the enhanced dissolution of Si and Al ions and subsequent formation of silicate and aluminate monopolymer. However, at relatively higher $SiO_2:Na_2O$ molar ratio (1.91) or lower alkalinity of the slurry, the geopolymeric reaction was similar to the hydration reaction of cement. The C-S-H gel can be formed along with the hydration reaction. As the reaction time was extended, the oligomers were formed to provide late-stage strength for green cement GPs.

4.4 BOND STRENGTH

The bond strength of GPs depends mostly on the ingredients and curing processes. It is important to consider the influence of these conditions on the bonding strength of GPMs [40–43]. Yet, no studies have been on the high-temperature–dependent bond strength of GPs.

4.4.1 Effect Calcium Content

Tanakorn et al. [17] studied the effect of calcium and C-S-H contents on the bonding strength of GP, where FA high calcium and OPC were used. OPC was replaced by

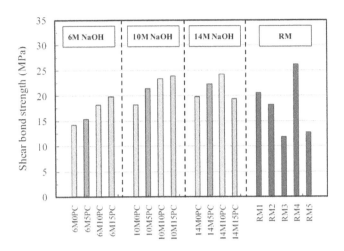

FIGURE 4.7 Shear bond strength of GPM or RM with interface line at 45° to the vertical [6].

FA in four batches such as 0%, 5%, 10% and 15% with varying molarity of NaOH. For all mixtures, the ratio of sodium silicate to sodium hydroxide was kept constant. Figure 4.7 shows the outcome of 45° slant shear load-carrying ability of Normal concrete (NC) substrate together with GPMs or RMs. The shear bond strengths were observed to increase when both OPC and NaOH contents were increased. This noticeable enhancement in shear bond strength was attributed to the augmentation of reaction products. This followed the earlier trends about the enhanced strength of FA-based GPs with high Ca content and co-existing C-S-H, C-A-S-H and N-A-S-H gels [40]. This enhancement of products at the boundary zone among Portland cement concrete (PCC) substrate and GPM was responsible for further increase in the strength at the contact zone [44]. Moreover, GPM with 15% of PC level and 14 M of NaOH displayed a small reduction in the shear bond strength. This reduction at high NaOH contents was also documented [10], which was ascribed to the accelerated dissolution of silica and alumina and subsequent inhibition of polycondensation reaction [45]. The observed lower strength of GPs at higher NaOH contents was due to the generation of excess OH ions that resulted in the precipitation of alumina-silicate gel at an early age [29]. Meanwhile, the dissolution of Ca was inhibited at higher NaOH contents and thereby resulted in less hydrated products.

The 45° slant shear load-carrying capacity of PCC substrate and RMs is in the range between 11.8 and 26.2 MPa compared to those of GPM in between 14.2 and 24.2 MPa. The mixes with 10 and 14 M of NaOH revealed considerably higher shear bonding strength compared with the average of RMs. This showed that GPM with OPC as an additive is indeed a potential alternative RM [6]. Phoongernkham et al. [17] reported the bond strength of GPMs. The slant shear ability of OPC concrete substrate (strength of 35 MPa) and GPP at 45° of interface line to the vertical were determined (Figure 4.8). The shear bond strengths were increased with the increase in GBFS content.

Hu et al. [27] reported the effect of GBFS substituted by MK-based GP (0% and 20%). It revealed that GPs possessed superior repairing attributes than OPC-based

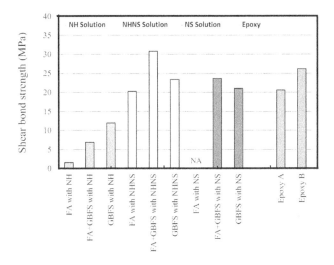

FIGURE 4.8 Shear bond strength between concrete substrate and GPP or epoxy with interface line at 45° to the vertical [17].

TABLE 4.5
Bond Strength and Failure Mode of the RMs [27]

Mix No.	Bond Strength (MPa)					Failure Mode				
	8 hours	1 day	3 days	7 days	28 days	8 hours	1 day	3 days	7 days	28 days
Cb	-	-	0.34	1.17	1.95	-	-	A	A	A
Gb	0.42	1.23	2.34	2.43	2.91	B	B	C	C	C
GSb	0.45	1.27	2.40	2.47	3.04	B	B	C	C	C

systems. Furthermore, the inclusion of steel slag can improve considerably the bond strength of GP specimens (Table 4.5). Among commercial RMs (Cb), GPM with 100% MK (Gb) and GP prepared with 20% GBFS (Gsb) are suited. Hawa et al. [23] reported the effect of PWA and rubber-wood ash substituted by MK on bond strength; PWA was found to increase the bond strength to Portland cement mortar quite dramatically.

4.4.2 Effect of Alkaline Activator Solution

Tanakorn et al. [6] reported the effect of NaOH concentration (6, 10 and 14 M) on the development of GPM bond strength. Use of GPM as RM revealed high-performance bonding at 10M of NaOH. The bond strength decreased with the decrease in NaOH molarity. Besides, the high molarity led to reduced bond strength of GPM containing high amount of calcium [6,17]. Phoongernkham et al. [17] reported the influence of solution types on the activation process of the waste materials. Three types of alkali activator were used such as NH, NS and NHNS. The NaOH mix with sodium silicate revealed high-performance bond strength than NaOH or sodium silicate as alkali solution.

4.4.3 Effect of Silicate-to-Aluminium Ratio

Zhang et al. [18] inspected the effect of Si:Al ratios on developed bonding strength of GPM. GPs with a low Si:Al ratio exhibited lesser bonding strength at ambient temperature but retained higher value at higher temperatures (Figure 4.9).

4.4.4 Effect of Solid-to-Liquid Ratio

The effect of varying solid-to-liquid ratio (0.5–1.1) on the bonding strength at ambient temperature was determined [18] (Figure 4.10). It was realized that very low (0.5) or very high (1.1) ratios of solid to liquid was disadvantageous to the bonding strength development of GPs, where bond failure was more susceptible. The two ratios of 0.6 and 0.8 produced excellent bond strength and optimal workability.

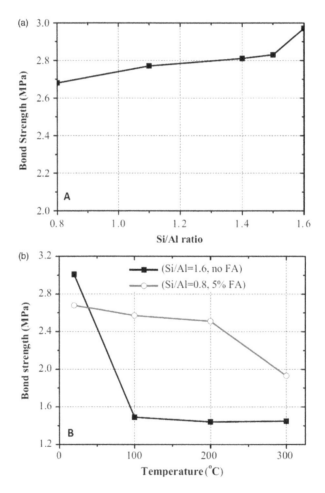

FIGURE 4.9 Effect of Si:Al ratio on the bond strength at (a) ambient temperature and (b) varying temperature [18].

Factors Effect on Geopolymer

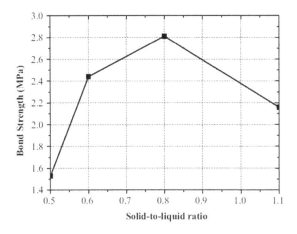

FIGURE 4.10 Effect of solid-to-liquid ratio on bond strength at ambient temperature [18].

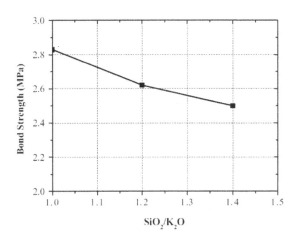

FIGURE 4.11 SiO_2:K_2O molar ratio–dependent bond strength at ambient temperature [18].

4.4.5 Effect of Curing Humidity

The influence of curing humidity on the bonding strength of GPs was studied in different environments for two types of mixes (A and B) with identical constituent materials. Group A was cured at 20°C in the presence of 90% humidity and Group B was cured at the same temperature but with lower humidity (50%). The bond strengths of A and B groups after 7 days of curing measured to be 2.83 and 2.43 MPa, respectively. The acquirement of relatively lower bond strength in Group B specimens was ascribed to the higher loss of water at lower levels of curing humidity [18].

4.4.6 Effect of SiO_2:K_2O Ratio

Zhang et al. [18] examined the SiO_2:K_2O (alkaline activator) molar ratio–dependent bond strength of GPs as depicted in Figure 4.11. The bond strength was observed to

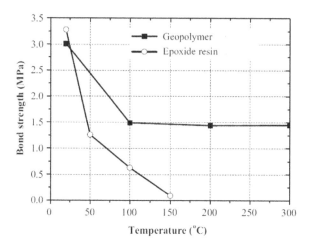

FIGURE 4.12 Temperature-dependent bond strength of various GPs [18].

increase slightly with decreasing SiO_2:K_2O ratio. The target ratio of SiO_2 to K_2O was obtained by dissolving KOH flakes in K_2SiO_3 solution. The reduction in the ratio of SiO_2 to K_2O caused an increased in the KOH contents in the activator solution.

4.4.7 BOND STRENGTH AT ELEVATED TEMPERATURES

Zhang et al. [18] examined the effect of increasing temperatures from 20°C to 300°C on the bonding strength of various kinds of GPs. They performed double shear tests on six types of GPs in the range of 20°C–300°C. Figure 4.12 compares the temperature-dependent bonding strength of studied GPs with that of epoxy resin. At ambient temperature, the prepared GPs exhibited somewhat weaker bonding strength than epoxy resin. Moreover, GPs showed superior bonding strength in the entire temperature region of 100°C–300°C.

4.5 FLEXURAL STRENGTH

Phoongernkham et al. [6] studied flexural strength of GPM used as RMs. High-content calcium materials (HFA) and OPC were used with different molarity of alkali activator solution. The bending stresses of PCC-notched beams filled with GPM or RM were observed to be larger at higher PC contents (Figure 4.13). This was primarily ascribed to the increase in the reaction products which in turn improved the bonding strength of GPMs. The GPM-filled notched beam (10% PC) activated with 14 M of NaOH solution produced outstanding bending stress (3.1 MPa) which corresponded to representing almost 85% enhancement from the baseline. The observed improved properties at higher NaOH activation were ascribed to the enhanced chemical reactions between NaOH and PCC substrates around the transition region. This test confirmed the suitability of PC incorporated with GPM as an alternative RM.

Dai et al. [26] determined the effect of SiO_2:Na_2O ratio on the flexural strength of GPM used as RM. The GPM prepared with higher SiO_2:Na_2O molar ratio revealed

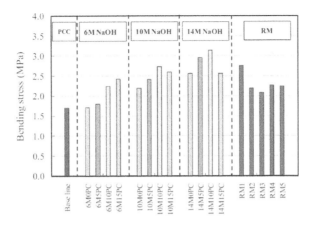

FIGURE 4.13 Bending stress of PCC-notched beam with filled GPM or RM [6].

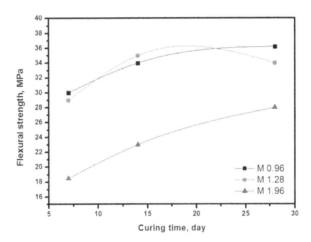

FIGURE 4.14 Flexural strength of GPM [26].

lower strength compared to other ratio (Figure 4.14). Torgal et al. [28] produced GPM as RMs by substituting TMW with 10% of calcium hydroxide. That GPM was found to possess much higher bond strength than current commercial repair products.

4.6 DRYING SHRINKAGE

Hawa et al. [23] studied the dry shrinkage of GPM as RMs. Three types of waste materials were used as binder to produce different mortar patches such as MK mixed with PWA or OPA. These GPMs were exposed to 30°C ± 2°C under 70% ± 5% of relative humidity for a prolonged period of up to 30 weeks. Figures 4.15–4.18 show the drying shrinkage capacity of all the samples. Drying shrinkage of the control samples was discerned to be lower than those with 10%, 20% and 30% of PWA (Figure 4.15). The average particle size of PWA being larger than MK correlated

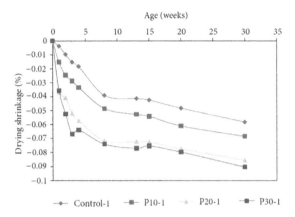

FIGURE 4.15 Drying shrinkages of GPMs cured at 80°C for 1 hour containing varied ratio of PWA [23].

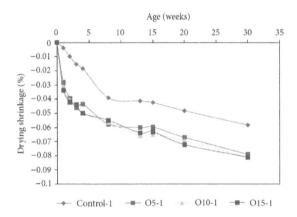

FIGURE 4.16 Drying shrinkages of GPMs cured at 80°C for 1 hour containing varied ratio of OPA [23].

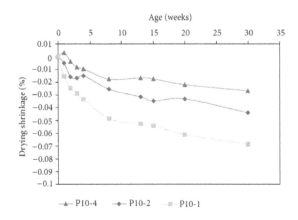

FIGURE 4.17 Drying shrinkage of GPMs containing 10% PWA cured at 80°C with different time [23].

Factors Effect on Geopolymer

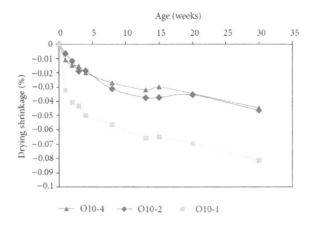

FIGURE 4.18 Drying shrinkage of GPMs containing 10% OPA cured at 80°C with different times [23].

negatively with the specific surface area. The GP reaction remained incomplete after heat curing at 80°C for 1 hour. It may be because of the excess water evaporation inability as samples were wrapped with polyvinyl sheets. Once the polyvinyl sheets were carved off, the pore water evaporated, which appeared easier for large pores corresponding to coarser particles. This is how the PWA content contributed and increased the drying shrinkage process relative to the control samples.

Furthermore, the presence of OPA increased the drying shrinkage (Figure 4.16). In the control samples with 5%, 10% or 15% OPA, drying shrinkage was observed to be rapid for 1- to 8-hour interval. The MK-based GP containing PWA or OPA revealed lower drying shrinkage capacity than those reported for slag mortar with sodium silicate, sodium hydroxide and sodium carbonate activators [46]. Figures 4.17 and 4.18 illustrate the drying shrinkage of GPM containing 10% PWA and 10% OPA with heat curing at 80°C for 1, 2 or 4 hours. All the samples exhibited similar behaviour in terms of longer curing and minimized drying shrinkage. The shrinkage of the samples containing 10% PWA for 4 hours curing displayed less shrinkage with shorter curing time. The use of water in geopolymerization reaction increased with higher curing time. The drying shrinkage of samples containing 10% OPA cured for 2 and 4 hours revealed similar trend at all ages up to 30 weeks. However, the drying shrinkage of the specimens cured for 1 hour was observed to be much higher at all ages than those with the longer cure periods. Again, the longer cure time left less water for evaporation at the expense of higher geopolymerization reactions.

4.7 ABRASION–EROSION RESISTANCE

Hu et al. [27] studied the AR of GPM used as RMs, where two types of GPM were prepared. One with MK 100% and the other MK replaced with 20% GBFS. The results were compared with cement as RMs. The AR of repairing specimens was determined at different ages (3, 7, 28, 56 and 90 days). Table 4.6 lists the measured values in terms of wear depth. The depth of wear was reduced with increasing curing ages. Comparing P_G with P_C, the p-values at 3, 7, 28, 56 and 90 days were observed to

TABLE 4.6
Depth of Grind Track of the RMs [27]

Mixture Number	P (mm)				
	3 days	7 days	28 days	56 days	90 days
Cement repair (P_C)	6.41	5.42	4.21	4.01	3.89
Geopolymeric repair (P_G)	3.32	3.03	2.98	2.87	2.76
Geopolymeric repair (P_{GS})	3.02	2.76	2.34	2.23	2.14

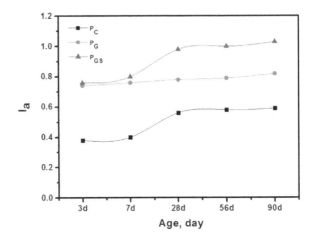

FIGURE 4.19 Age-dependent grades of AR of the RMs [27].

decrease to 48%, 44%, 29%, 28% and 29%, respectively. Comparing P_{GS} with P_G, the p-values at 3, 7, 28, 56 and 90 days were found to reduce to 9%, 8.9%, 21.4%, 22.2% and 22.5%, respectively. It was evident that the depth of wear for GPM specimens was slighter compared to the OPC-based repairing substances. Thus, the AR for GPs as repairing product was declared to be superior. This declining nature of the P_G and P_{GS} values than P_C for 3–28 days was similar to the variation of compressive strengths. It was majorly attributed to the alteration in the structural density and its influence on the AR of the repairing GP specimens, where the GP might form a denser structure compared to the OPC-based substances.

Figure 4.19 displays the calculated age-dependent ranking of AR (I_a) of the repairing GP specimens, which was found to increase as the age increased. Furthermore, the value of I_a for the geopolymeric RM was observed to be higher compared to the cement RM. Yet, I_a for the GR was lower than the steel slag.

4.8 MICROSTRUCTURES

Phoongernkham et al. [17] investigated the influence of alkali solution types on the morphology of FA-GBFS GP. The scanning electron microscope (SEM) images revealed the fractured surfaces of GPPs (Figure 4.20). Generally, the FA pastes possessed wobbly matrix. However, with NH and NS activation, FA pastes are less dense

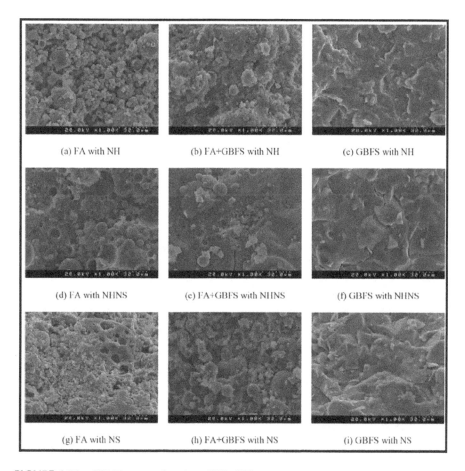

FIGURE 4.20 SEM images of various GPPs [17].

and still have loose-fitting network. Yet, FA paste activated with NHNS revealed several non-reacted and/or incompletely reacted FA particles surrounded in an uninterrupted matrix. The NHNS-activated FA paste became somewhat denser compared to the pastes with NH and NS activation. This supported the earlier observation, which confirmed that using NH plus NS it was possible to speed up the process of geopolymerization [36]. Furthermore, FA plus GBFS mix revealed substantial distinction than FA pastes. The FESEM images displayed fewer non-reacted FA particles and the matrices were denser compared to FA pastes. FA pastes revealed comparatively slower rate of reaction at ambient temperature and curing temperature variation enhanced the development of strength [8,47].

An increase in the content of GBFS was found to accelerate the rate of reaction where the exothermal reaction between GBFS and alkaline solution generated excess heat and led to the formation of extra C-S-H and C-A-S-H gels, thereby improving the strength [38,48]. The FA plus GBFS paste activated with NHNS solution was denser compared to the paste activated with NH and NS solution. The NHNS activation was better in terms of accelerated geopolymerization of FA-GBFS–based GPP product than the one with single activator as reported [49]. Relatively higher strength

(114.5 MPa) at 28 days was achieved for FA plus GBFS paste mix activated with the solution of NHNS. The matrix of GBFS paste was highly dense and homogeneous due to the formation of extra C-S-H gels [38] which led to the microstructure modification of the mixes. Furthermore, mixes with Na_2SiO_3 (NHNS in Figure 4.20f and NS in Figure 4.20i) revealed higher density than the one prepared using NH (Figure 4.20c). The presence of SiO_2 in the pastes promoted its reaction with Ca and contributed to the formation of calcium silicate (Ca_2SiO_4) products.

The SEM data of Phoongernkham et al. [6] revealed the existence of fracture interfaces between PCC substrate and GPM or RM (Figure 4.21). The fractured

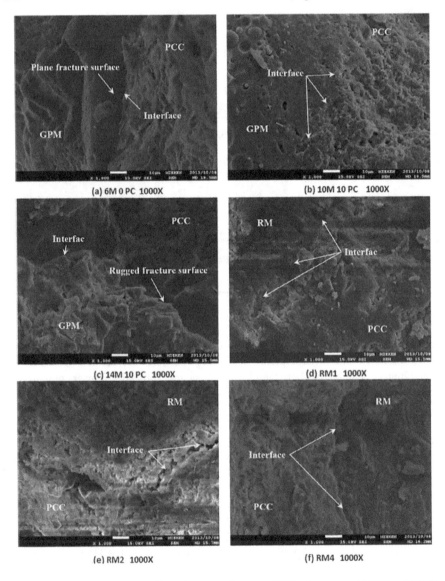

FIGURE 4.21 SEM images of interface zone between PCC substrate and GPM or RM [6].

interfaces of the mortar activated with low contents of NaOH and without containing OPC (6M0PC) (Figure 4.21a) showed comparatively planer surfaces representing their cleaner partitioning and weak bonding. This attained both small bending stress and shear bonding strength of the mix that contained tiny amount of NaOH and no OPC (6M0PC). The fractured interfaces for both mixes with high amount of NaOH and OPC mortar (10M10PC) (Figure 4.21b) revealed an intact bonding surface. This observation of insignificant gap between the two bonding interfaces was attributed to the passage of cracks through PCC substrate and GPM interface. The 14M10PC mix (Figure 4.21c) displayed an extremely uneven or rugged fractured interface without any noticeable planer fractured region. This indicated an excellent bonding at the interfacial region. These findings verified the enhancement of the shear bonding strength and the bending stress for mixes containing large amount of NaOH and OPC mortar. Figure 4.21d and e shows the associated monolithic failure mode.

The augmentation in the bonding strength was acknowledged by Pacheco-Torgal et al. [35]. Shi et al. [49] reported the occurrence of the elevated tensile strength ratio and improved the bonding of alkali solution–activated binder than conventional OPC. Besides, it was found that cementitious nature of recycled concrete can be improved by the activation of alkali solution and $Ca(OH)_2$ in the remaining paste [50]. It is known that the richness of Si^{4+} and Al^{3+} ion contents in GP allows the efficient reaction with $Ca(OH)_2$ at the PCC substrate surface and thereby leads to the enhanced strength development at the contact zone. The balancing of the negatively charged Al^{3+} ions with increasing Ca^{2+} ion contents could enhance the reaction products at the interface region between the PCC and GP mix. Consequently, a dense interfacial region is formed which contributes to the development of elevated strength of GP.

The SEM images (Figure 4.21d) of the prism sample (RM1) having elevated shear bond and large bending strengths revealed moderately dense surface in containing a tiny gap. The SEM images (Figure 4.21e) of the prism sample (RM2) displayed a noticeable gap at the interface demonstrating a moderately lower bonding. This supported the observed failure mode and the formation of cracks in RM interface while the PCC substrate was intact. The SEM images (Figure 4.21f) of the prism sample (RM4) showed only a tiny gap at the interface. The SEM images of RM1 and RM4 demonstrated superior bonding strength of the repairing product and matched well with the monolithic failure mode.

4.9 FAILURE MODE AND INTERFACE ZONE

Phoongernkham et al. [6] investigated the failure modes of slant shear prisms (Figures 4.22 and 4.23) and identified two failure patterns. The first one was in the GPM, where the cracks were generated in the interface while the PCC substrate remained reasonably intact (Figure 4.22). This occurred at low NaOH concentration without PC and low PC content (6M0PC and 6M5PC mixes) GPMs. For other mixes with relatively high strengths, high NaOH and high PC such as 10M10PC and 14M10PC mixes, and the slant shear bond prisms failed in the monolithic mode. In menace, cracks were formed in both sections of GPM and PCC substrate. This clearly indicated GPMs' higher resistance to cracking and superior bonding at the

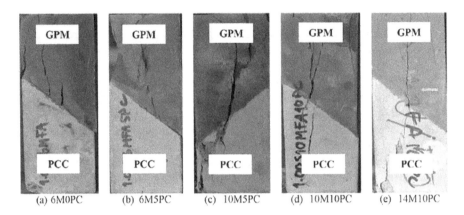

FIGURE 4.22 Fracture surface between PCC substrate and GPM [6].

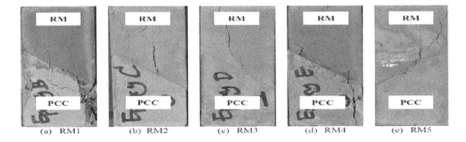

FIGURE 4.23 Fracture surface between PCC substrate and RM [6].

interface. The observed monolithic kind of failure was attributed to the propagation of cracks through the slant plane.

Two kinds of failure patterns were evidenced for the RM. In the first kind, the cracks were generated in the RM and at the interface, while the PCC substrate remained moderately intact (Figure 4.23; prisms having RM2, RM3 and RM5). These findings agreed well with the weak shear bonding strengths of the prisms (RM2, RM3 and RM5). Figure 4.23a and d displays the monolithic mode failures of the prisms with RM1 and RM4. Yet again, comparatively elevated resistance towards the cracking of RM1 and RM4 and the high bonding at the interface were observed. These findings tallied with the large shear bonding strengths of the prisms with RM1 and RM4.

4.10 SUMMARY

Based on the in-depth analyses of the obtained results, the following conclusions can be drawn:

i. The GP workability as RMs was found to decrease with increasing the alkaline activator solution molarity, sodium silicate to sodium hydroxide and calcium oxide content. The shorter setting time was observed with the mixtures containing high amount of GBFS.

ii. The engineering properties of GPs were highly influenced by alkaline activator solution molarity and calcium content. The lower strength was obtained with the specimens prepared with low molarity and calcium content.
iii. The GP performance as RMs was highly influenced by binder to fine aggregate, solution content, solution modulus and curing regime. An increase in the content of GBFS was found to accelerate the rate of reaction where the exothermal reaction between GBFS and alkaline solution generated excess heat and led to the formation of extra C-S-H and C-A-S-H gels, thereby improved the strength and bond zone. This clearly indicated GPMs' higher resistance to cracking and superior bonding at the interface.
iv. Increasing calcium-to-silica ratio led to increase the drying shrinkage and effect negatively on GP as concrete surface RMs.
v. From microstructure results, the fractured interfaces of the GP activated with low contents of NaOH and without containing any calcium resources showed comparatively planer surfaces representing their cleaner partitioning and weak bonding.

REFERENCES

1. Alanazi, H., et al., Bond strength of PCC pavement repairs using metakaolin-based geopolymer mortar. *Cement and Concrete Composites*, 2016. **65**: pp. 75–82.
2. Pacheco-Torgal, F., et al., *Handbook of alkali-activated cements, mortars and concretes*. 2014: Elsevier.
3. Haneefa, K.M., M. Santhanam, and F. Parida, Performance characterization of geopolymer composites for hot sodium exposed sacrificial layer in fast breeder reactors. *Nuclear Engineering and Design*, 2013. **265**: pp. 542–553.
4. Kumar, S., R. Kumar, and S. Mehrotra, Influence of granulated blast furnace slag on the reaction, structure and properties of fly ash based geopolymer. *Journal of Materials Science*, 2010. **45**(3): pp. 607–615.
5. Sukmak, P., S. Horpibulsuk, and S.-L. Shen, Strength development in clay–fly ash geopolymer. *Construction and Building Materials*, 2013. **40**: pp. 566–574.
6. Phoo-ngernkham, T., et al., Effects of sodium hydroxide and sodium silicate solutions on compressive and shear bond strengths of FA–GBFS geopolymer. *Construction and Building Materials*, 2015. **91**: pp. 1–8.
7. Chindaprasirt, P., et al., High-strength geopolymer using fine high-calcium fly ash. *Journal of Materials in Civil Engineering*, 2010. **23**(3): pp. 264–270.
8. Pangdaeng, S., et al., Influence of curing conditions on properties of high calcium fly ash geopolymer containing Portland cement as additive. *Materials & Design*, 2014. **53**: pp. 269–274.
9. Phoo-ngernkham, T., et al., The effect of adding nano-SiO_2 and nano-Al_2O_3 on properties of high calcium fly ash geopolymer cured at ambient temperature. *Materials & Design*, 2014. **55**: pp. 58–65.
10. Somna, K., et al., NaOH-activated ground fly ash geopolymer cured at ambient temperature. *Fuel*, 2011. **90**(6): pp. 2118–2124.
11. Islam, A., et al., The development of compressive strength of ground granulated blast furnace slag-palm oil fuel ash-fly ash based geopolymer mortar. *Materials & Design*, 2014. **56**: pp. 833–841.
12. Huseien, G.F., et al., Effects of POFA replaced with FA on durability properties of GBFS included alkali activated mortars. *Construction and Building Materials*, 2018. **175**: pp. 174–186.

13. Nath, P. and P.K. Sarker, Use of OPC to improve setting and early strength properties of low calcium fly ash geopolymer concrete cured at room temperature. *Cement and Concrete Composites*, 2015. **55**: pp. 205–214.
14. Abdalla, H. and B. Karihaloo, Determination of size-independent specific fracture energy of concrete from three-point bend and wedge splitting tests. *Magazine of Concrete Research*, 2003. **55**(2): pp. 133–142.
15. Phoo-ngernkham, T., et al., Properties of high calcium fly ash geopolymer pastes with Portland cement as an additive. *International Journal of Minerals, Metallurgy, and Materials*, 2013. **20**(2): pp. 214–220.
16. Suwan, T. and M. Fan, Influence of OPC replacement and manufacturing procedures on the properties of self-cured geopolymer. *Construction and Building Materials*, 2014. **73**: pp. 551–561.
17. Phoo-ngernkham, T., et al., High calcium fly ash geopolymer mortar containing Portland cement for use as repair material. *Construction and Building Materials*, 2015. **98**: pp. 482–488.
18. Zhang, H.Y., et al., Characterizing the bond strength of geopolymers at ambient and elevated temperatures. *Cement and Concrete Composites*, 2015. **58**: pp. 40–49.
19. Kong, D.L., J.G. Sanjayan, and K. Sagoe-Crentsil, Comparative performance of geopolymers made with metakaolin and fly ash after exposure to elevated temperatures. *Cement and Concrete Research*, 2007. **37**(12): pp. 1583–1589.
20. Perera, D., et al., Influence of curing schedule on the integrity of geopolymers. *Journal of Materials Science*, 2007. **42**(9): pp. 3099–3106.
21. Kuenzel, C., et al., Ambient temperature drying shrinkage and cracking in metakaolin-based geopolymers. *Journal of the American Ceramic Society*, 2012. **95**(10): pp. 3270–3277.
22. He, J., et al., Geopolymer-based smart adhesives for infrastructure health monitoring: concept and feasibility. *Journal of Materials in Civil Engineering*, 2010. **23**(2): pp. 100–109.
23. Hawa, A., et al., Development and performance evaluation of very high early strength geopolymer for rapid road repair. *Advances in Materials Science and Engineering*, 2013. **2013**.
24. Vasconcelos, E., et al. Concrete retrofitting using CFRP and geopolymer mortars. in *Materials science forum*. 2013: Trans Tech Publ.
25. Moura, D., et al. Concrete repair with geopolymeric mortars: influence of mortars composition on their workability and mechanical strength. in *VI international materials symposium (materials 2011) University of Minho, Guimarães, Portugal*. 2011: pp. 1–16.
26. Dai, Y., et al., *A study on application of geopolymeric green cement*. National Taipei University of Technology, Taiwan. 2013: pp. 1–8.
27. Hu, S., et al., Bonding and abrasion resistance of geopolymeric repair material made with steel slag. *Cement and Concrete Composites*, 2008. **30**(3): pp. 239–244.
28. Torgal, F.P., J. Gomes, and S. Jalali, Bond strength between concrete substance and repair materials: comparisons between tungsten mine waste geopolymeric binder versus current commercial repair products. in *Seventh International Congress on Advances in Civil Engineering*, Istanbul, Turkey. 2006: pp. 1–10.
29. Lee, W. and J. Van Deventer, The effects of inorganic salt contamination on the strength and durability of geopolymers. *Colloids and Surfaces A: Physicochemical and Engineering Aspects*, 2002. **211**(2): pp. 115–126.
30. Chindaprasirt, P., et al., Effect of SiO_2 and Al_2O_3 on the setting and hardening of high calcium fly ash-based geopolymer systems. *Journal of Materials Science*, 2012. **47**(12): pp. 4876–4883.
31. Van Jaarsveld, J., J. Van Deventer, and G. Lukey, The characterisation of source materials in fly ash-based geopolymers. *Materials Letters*, 2003. **57**(7): pp. 1272–1280.

32. Winnefeld, F., et al., Assessment of phase formation in alkali activated low and high calcium fly ashes in building materials. *Construction and Building Materials*, 2010. **24**(6): pp. 1086–1093.
33. Temuujin, J., A. Van Riessen, and R. Williams, Influence of calcium compounds on the mechanical properties of fly ash geopolymer pastes. *Journal of Hazardous Materials*, 2009. **167**(1): pp. 82–88.
34. Hawa, A., D. Tonnayopas, and W. Prachasaree, Performance evaluation and microstructure characterization of metakaolin-based geopolymer containing oil palm ash. *The Scientific World Journal*, 2013. **2013**: pp. 1–10.
35. Temuujin, J., R. Williams, and A. Van Riessen, Effect of mechanical activation of fly ash on the properties of geopolymer cured at ambient temperature. *Journal of Materials Processing Technology*, 2009. **209**(12): pp. 5276–5280.
36. Rattanasak, U. and P. Chindaprasirt, Influence of NaOH solution on the synthesis of fly ash geopolymer. *Minerals Engineering*, 2009. **22**(12): pp. 1073–1078.
37. Rashad, A., et al., Hydration and properties of sodium sulfate activated slag. *Cement and Concrete Composites*, 2013. **37**: pp. 20–29.
38. Ismail, I., et al., Modification of phase evolution in alkali-activated blast furnace slag by the incorporation of fly ash. *Cement and Concrete Composites*, 2014. **45**: pp. 125–135.
39. Duxson, P., et al., The role of inorganic polymer technology in the development of 'green concrete'. *Cement and Concrete Research*, 2007. **37**(12): pp. 1590–1597.
40. Dombrowski, K., A. Buchwald, and M. Weil, The influence of calcium content on the structure and thermal performance of fly ash based geopolymers. *Journal of Materials Science*, 2007. **42**(9): pp. 3033–3043.
41. Toutanji, H. and Y. Deng, Comparison between organic and inorganic matrices for RC beams strengthened with carbon fiber sheets. *Journal of Composites for Construction*, 2007. **11**(5): pp. 507–513.
42. Kurtz, S. and P. Balaguru, Comparison of inorganic and organic matrices for strengthening of RC beams with carbon sheets. *Journal of Structural Engineering*, 2001. **127**(1): pp. 35–42.
43. Toutanji, H., L. Zhao, and Y. Zhang, Flexural behavior of reinforced concrete beams externally strengthened with CFRP sheets bonded with an inorganic matrix. *Engineering Structures*, 2006. **28**(4): pp. 557–566.
44. Pacheco-Torgal, F., J. Castro-Gomes, and S. Jalali, Adhesion characterization of tungsten mine waste geopolymeric binder. Influence of OPC concrete substrate surface treatment. *Construction and Building Materials*, 2008. **22**(3): pp. 154–161.
45. Zuhua, Z., et al., Role of water in the synthesis of calcined kaolin-based geopolymer. *Applied Clay Science*, 2009. **43**(2): pp. 218–223.
46. Atiş, C.D., et al., Influence of activator on the strength and drying shrinkage of alkali-activated slag mortar. *Construction and Building Materials*, 2009. **23**(1): pp. 548–555.
47. Wongpa, J., et al., Compressive strength, modulus of elasticity, and water permeability of inorganic polymer concrete. *Materials & Design*, 2010. **31**(10): pp. 4748–4754.
48. Li, C., H. Sun, and L. Li, A review: the comparison between alkali-activated slag (Si+Ca) and metakaolin (Si+Al) cements. *Cement and Concrete Research*, 2010. **40**(9): pp. 1341–1349.
49. Shi, X., et al., Mechanical properties and microstructure analysis of fly ash geopolymeric recycled concrete. *Journal of Hazardous Materials*, 2012. **237**: pp. 20–29.
50. Achtemichuk, S., et al., The utilization of recycled concrete aggregate to produce controlled low-strength materials without using Portland cement. *Cement and Concrete Composites*, 2009. **31**(8): pp. 564–569.

5 Performance Criteria of Geopolymer as Repair Materials

5.1 INTRODUCTION

Repair or rehabilitation is the major concern regarding several deteriorated concrete structures [1–4]. Such repairs of concrete structures are necessary to assure their service lifetimes. Moreover, they must be completed in a short time for public convenience [3,5]. Over the years, several repair materials are developed for concrete structures including cement-based materials, polymers and latex [5,6]. Recently, geopolymer mortars (GPMs) revealed tremendous prospects towards emergency repairs and coating [7–11]. The notion of geopolymer (GP) was first introduced in the late 1970s, where aluminosilicate binders was activated by alkali solution to describe a family of GP [12]. The formation of GP was based on the reaction between the two parts of materials such as the alkali activator and the reactive aluminosilicate precursor (mainly MK) [13]. The GP based on alkali activation of MK became attractive not only because of its excellent thermal stability (better than conventional polymer material) but also due to its comparable mechanical properties to cement. Nowadays, GPMs are considered as a green alternative to Portland cement [14].

Despite the wide usage of fly ash (FA) and slag as two major materials in commercial GP products, MK emerged as most promising future feedstock materials for GP. It is needless to mention that MK possesses more consistent chemical compositions than FA and slag. Thus, it results in more reliable and predictable products that are suitable for repairable construction materials. Actually, both FA and slag are not abundant in many countries because of their effective usage in the manufacture of blending cements and concrete [15,16]. On top of this, the cost and technical challenges of supply chain limit their widespread usage [17]. Thus, the use of MK (together with other Al- and Si-bearing minerals) as a raw material appears more prospective and practicable [13].

Recently, intensive researchers are carried out on GPMs to understand the mechanism of geopolymerization and optimization of the product for achieving improved strength. Bernal et al. [18] studied the evolution of the binder structure of sodium silicate (NS)–activated slag-MK blends, where the effect of MK addition on the final strength of the binder is examined. Silva and Sagoe-Crenstil [19] determined the effect of different ratios of Al_2O_3 and SiO_2 on the setting and the hardening of the GP system. It is acknowledged that this ratio indeed affects the setting time and the final strength of the achieved GP. Chindaprasirt et al. [20] investigated the influence of $SiO_2:Al_2O_3$ on the setting time, the workability and the final strength of the GP system and found the best ratio for GP binder ($SiO_2:Al_2O_3$) around 2.87–4.79.

DOI: 10.1201/9781003173618-5

In the past few years, several researchers [21–25] have attempted to utilize GP as a repair material by testing their slant shear, pull-out and direct shear. Hu et al. [21] studied the bond strength between mortar substrate and GP in sandwich specimens. Geopolymer exhibited higher bonding strength than that of comparable Portland cement mixture. Pacheco-Torgal et al. [22] determined the bond strength between concrete substrate and GPM that were produced from tungsten mine waste containing calcium hydroxide. Phoongernkham et al. [23,24] examined the effect of molarity of sodium hydroxide (NH), NS content and calcium-to-silicate ratio on shear bond of GPM as repair material. They found that geopolymeric binders possess very high bond strength even at an early age as compared to commercial repair products.

Considering these interesting attributes of MK, we inspected the effect of MK on the early bond strength of GPM to realize its potential as a repair material [26]. The so-called bond strength between a repair material and an existing concrete being one of the most critical factors impacting the repair durability was evaluated using a splitting tensile and slant shear test of the produced GPMs. Tests were systematically conducted to characterize the bond strength between fabricated GPMs and mortar substrate. The results were analysed, discussed and compared with commercially available repair materials.

5.2 GEOPOLYMER BINDER

To assess the performance of GP as a concrete repair material, binary blend was used that included metakaolin (MK) and ground blast furnace slag (GBFS). MK is mainly characterized as a source of aluminosilicate for the preparation of GP. In this experiment, we followed the earlier procedures to prepare MK from kaolin through dihydroxylation in the furnace, where kaolin was calcined at 750°C for 6 hours [27]. Kaolin powder (grade KM40) was purchased from the kaolin Malaysia SDN (Puchong, Selangor).

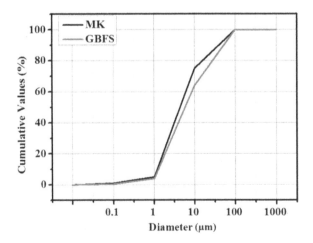

FIGURE 5.1 Cumulative particle size distribution of MK and GBFS.

Performance Criteria of Geopolymer 79

MK has distinctive white colour close to that of parent kaolin material. Moreover, the appearance of kaolin has changed from pure white to floral white after dehydroxylization process. Malvern Mastersizer micro-particle size analyser was used to determine the particle size distribution of MK. Figure 5.1 illustrates the result of particle size analysis. The particle of MK cut at 1 μm was found to be characteristically below 100 μm, where 75% of MK had passed through 10 μm. The scanning electron microscope (SEM) images of MK sample revealed irregular platy. Angular-shaped particles were observed to be closely packed in lumps with disorder arrangements (Figure 5.2). The tiny lumps observable in their physical form may be noticed in the micrograph appearing as stacks of layers of MK sheets. The X-ray diffraction (XRD) patterns of MK (Figure 5.3) showed a broad hunch between 9.8°C and 28°C and a sharp crystalline peak at 26.8°C, which are allocated to the presence of amorphous structure of quartz (SiO_2), mullite ($Al_6Si_2O_{13}$), andalusite (Al_2SiO_3), calcium oxide (CaO), magnesium silicate ($MgSiO_2$) and aluminium–magnesium (AlMg) crystalline phases. The quartz is generally known to be unreactive while the presence of

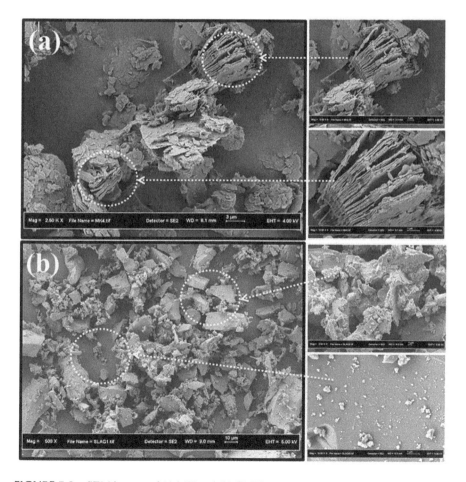

FIGURE 5.2 SEM images of (a) MK and (b) GBFS.

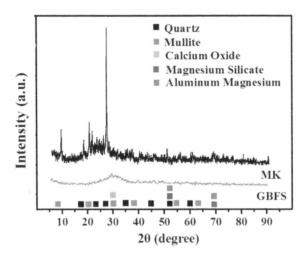

FIGURE 5.3 XRD patterns of MK and GBFS.

TABLE 5.1
Chemical Compositions of GBFS and MK (mass%)

Materials	SiO$_2$	Al$_2$O$_3$	CaO	MgO	Fe$_2$O$_3$	Na$_2$O	K$_2$O	TiO$_2$	MnO	P$_2$O$_5$	LOI
MK	52.22	41.41	0.08	0.26	0.49	0.01	1.73	0.01	0.01	0.13	1.66
GBFS	30.53	13.67	46.02	5.09	0.33	0.24	0.36	-	-	0.01	0.22

muscovite (KAl$_2$(Si$_3$Al)O$_{10}$(OH,F)$_{12}$) which is impure from the client is consumed during synthesis [27].

The chemical compositions of MK were determined using X-ray fluorescence (XRF) spectroscopy. The XRF results revealed that the major constituents of MK are silicon oxide (SiO$_2$) and alumina oxide (Al$_2$O$_3$). Other components include ferric oxide (Fe$_2$O$_3$), calcium oxide, magnesium oxide and potassium oxide. The typical chemical composition of MK is depicted in Table 5.1. MK should meet the requirements of ASTM C618 [28] (SiO$_2$ plus Al$_2$O$_3$ and Fe$_2$O$_3$) more than 85%.

The cement-free binder is made using GBFS as one of the resource materials, which is collected from Ipoh (Malaysia). GBFS possesses both cementitious and pozzolanic properties and is considered to be different from other supplementary cementitious materials. GBFS develops its own hydraulic reaction when mixed with water and is off-white in colour. The result of particle size showed that more than 60% particle has size lower than 10 μm. Figure 5.2 displays the SEM images of GBFS, which consisted of irregular, angular and spherical particles with a smooth surface. The XRD pattern of as-received GBFS revealed mainly the amorphous phase with a small amount of magnetite. The GBFS comprising calcium silicate and alumina (about 90%) meets the requirement of pozzolanic material [28].

5.3 GEOPOLYMER MIX DESIGN

In the preparation of GP specimens, two-part alkaline activator solution was used. The alkaline solution used in this experiment was a mixture of NS and NH (purity 98%). These were used to activate the alumina and silica in MK and GBFS. The NS solution was composed of SiO_2 (29.5 mass%), Na_2O (14.70 mass%) and H_2O (55.80 mass%). These chemicals were purchased from QREC (ASIA) SDN BHD (Malaysia). A different amount of pellets was dissolved in water to prepare NH solution of various molar concentrations (10, 14, 16 and 18 M). The solution was left for 24 hours to cool down, then it was added to NS solution to prepare the final alkaline solution with different mass ratios of SiO_2:Na_2O as listed in Table 5.2. The ratio of NS to NH was fixed for all mixtures.

Fine aggregate, naturally occurring siliceous river sand was used to make all mortar specimens. The sand was dried in the oven at 60°C for 24 hours for controlling the moisture content. The sand was graded to conform to ASTM C33 [29] standard specification (Figure 5.4). Fineness modulus of the aggregate and specific gravity were discerned to be 2.9 and 2.6, respectively. To increase the workability of

TABLE 5.2
Compositions of Alkaline Solution

Alkaline Solution	NaOH Solution (NH)			Na_2SiO_3 Solution (NS)			NS:NH	SiO_2:Na_2O
	Molarity M	Na_2O Mass%	H_2O Mass%	SiO_2 Mass%	Na_2O Mass%	H_2O Mass%	Mass%	Mass%
S1	10	28.5	71.5	29.5	14.7	55.8	3.0	1.26
S2	14	35.9	64.1	29.5	14.7	55.8	3.0	1.16
S3	16	39.2	60.8	29.5	14.7	55.8	3.0	1.12
S4	18	42.1	57.9	29.5	14.7	55.8	3.0	1.08

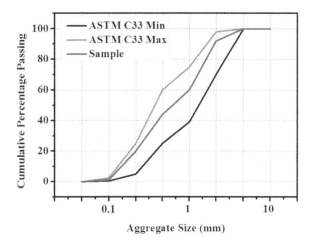

FIGURE 5.4 Particle size analysis of fine aggregate.

GPM, super-plasticizer (SP) type (Sika Visco Crete-3430) was used. The utilization of viscosity-modifying admixture provides more possibilities of controlling segregation (stability) and homogeneity of the mix. The amount of SP was kept fixed for all mixtures with 3% of the binder.

Figure 5.5 displays the four stages of mix proportions of GPMs. In the first stage, the effect of MK replacement by an amount of 0, 5, 10 and 15 mass% on GBFS was inspected, where other parameters were kept constant for all mixtures to select the optimum ratio of MK substituted by GBFS (Table 5.3). The high early strength after 24 hours is the criteria to select the optimum mixture at all stages. In the second stage, the impact

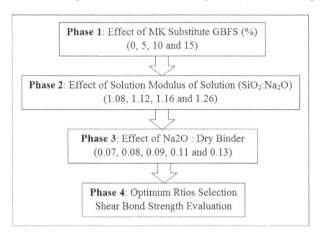

FIGURE 5.5 Mixture procedures.

TABLE 5.3
Mix Proportions of GPMs

Phase	Binder Mass% GBFS	Binder Mass% MK	Alkaline Solution Type[a]	S:B Mass%	SP Mass%	B:A Mass%	Na$_2$O:Dry Binder Mass%	H$_2$O:Dry Binder Mass%
1	100	0	S3	0.65	0.03	0.9	0.135	0.37
	95	5						
	90	10						
	85	15						
2	95	5	S4	0.65	0.03	0.9	0.14	0.37
			S2					0.38
			S1					0.39
3	95	5	S2	0.55			0.11	0.32
				0.45	0.03	0.9	0.09	0.26
				0.40				
				0.35				

[a] From Table 5.2.

of varying alkaline solution concentration ratio (SiO_2:Na_2O) of 1.26, 1.16, 1.12 and 1.08 on the strength of prepared mortars (optimum MK replaced by GBFS from Stage 1) was examined, where the alkaline solution-to-binder (S:B) ratio (0.65) for all mixtures were kept constant. In the third stage, the effect of various (7%, 8%, 9%, 11% and 13%) Na_2O:dry binder ratios on the strength of prepared mortars (the optimum percentage of MK substituted by GBFS and the optimum alkaline solution obtained from Phase 1 and 2, respectively) was determined. The NS-to-NH ratio of 3.0, binder-to-fine aggregate ratio (B:A) of 0.90 and amount of SP (3%) were kept constant in all three stages (Table 5.3) for the mixtures 1, 2 and 3. In the fourth stage, the optimum ratios from the previous three stages are selected to evaluate the bond strength of GPM. The results are compared with normal Ordinary Portland Cement (OPC) mortar as a control sample (NC).

The OPC is obtained from Holcim Cement Manufacturing Company (Malaysia) conforming to ASTM C150 standard specification for Portland cement of 2009. This was used to prepare cement mortar samples. The calcium oxide (CaO) content was found to be around 62.7 mass%. CaO is considered to be the important composition of OPC as far as the hydration process is concerned. Actually, the CaO content in OPC must be very high. The control sample was made of OPC, fine aggregates and water. Fine aggregate-to-cement ratio by mass of 3:1 was considered. The fine aggregate was kept in a saturated surface under dry condition. The water-to-cement (w:c) ratio was set at 0.48.

Present GPMs were prepared by mixing MK with GBFS over a period of 4 minutes at dry condition to achieve a homogeneous mixture of fine aggregates. Then, the acquired mixture was activated by adding the alkaline solution to obtain a thorough mixed mortar cast into 50 mm cube moulds. The flow of the fresh GPM was measured to examine the effect of different parameters on the workability and setting time. The casting was performed in two layers, where each layer was compacted with a vibration table for 15 seconds. The samples were left for 24 hours after casting and before opening the moulds and curing at ambient temperature (27°C and 75% relative humidity). They were tested for 1, 3, 7 and 28 days to evaluate the compressive strength (according to ASTM 109 [30]) and other mechanical properties. Table 5.3 depicts the achieved three different phases of mixtures.

5.4 WORKABILITY PERFORMANCE

Flowability of geopolymer mortar was measured using a flow table method modified from ASTM C230, "Standard Specification for Flow Table for Use in Tests of Hydraulic Cement". The flow table provides an efficient means of determining the flow of cement pastes and hydraulic cement mortars. Vicat needle was used to measure the setting time (according to ASTM C191 standard), where the specimen was placed on the Vicat apparatus to measure the initial and final setting time.

Two different tests such as flow and setting time were performed on fresh state mortar. Figures 5.6 and 5.7 illustrate the effect of MK replaced by GBFS on the flow and the setting time of GPMs, respectively. The flow of the mortar was found to increase with increasing percentage of MK replaced by GBFS, which was attributed to the differences in the physical properties and chemical reactions of the mixtures. Furthermore, with the reduction of the GBFS content, the number of angular particles was reduced and helped to improve the flowability of the mortar mixture. Also, the calcium content was decreased with increasing MK content. Furthermore, the silicate and aluminium content was increased. More silicate content was useful for improving the flowability of the mortar. The fineness in the particle size of MK (75% of them less than 10 μm) has also contributed towards the increase of workability of the mortar. Similar trends were also reported earlier [31], where the particle size played an important role and affected the dissolution and flow of the mortar. On top of this, admixing of MK and GBFS produced a slow setting and enhanced the workability (Figure 5.6). This observation is supported by the findings of al Majidi et al. [32].

Figure 5.7 demonstrates the effect of MK replaced by GBFS on the setting time of GPM. It was observed that a decrease in the calcium content led to an enhancement in the initial and final setting times as reported elsewhere [23,31,33]. Furthermore, an increase in the MK content has enhanced the SiO_2 and Al_2O_3 concentrations, thereby improved the setting time [20]. The rate of setting time was increased significantly as indicated by the substantial difference in the initial setting time. The difference between initial and final setting time was also increased with the reduction of GBFS content in the mortar. This finding also supported the fact that higher the GBFS

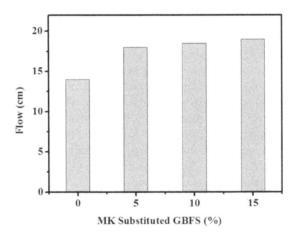

FIGURE 5.6 Impact of MK substituted by GBFS on the flow of GPMs.

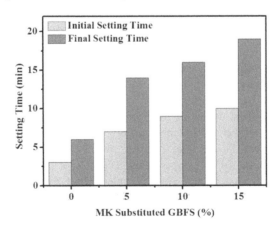

FIGURE 5.7 Impact of MK substituted by GBFS on the setting time of GPMs.

content in the mortar, the quicker the rate of setting [34,35]. Thus, it is established that MK as a part of the binary-blended binder is greatly effective to decelerate the setting time of GPMs under ambient conditions.

Figure 5.8 shows the effects of alkaline solution molarity variation on the flow of GPMs. The flow of the mortar was found to be higher at lower Na_2O content. The flow of GPMs was reduced from 23 to 15 cm as the Na_2O amount was enhanced with NaOH molarity increase from 10 to 18 M. Figure 5.8 displays the solution molarity–dependent variation on the workability of GPMs. An increase in the Na_2O content was observed to diminish the flow. Besides, an increase in the Na_2O content in the alkaline solution led to an increase in the sodium ion (Na^+) content and reduced the (SiO_2:Na_2O) ratio (Table 5.2). This led to an increase in the heat released, thereby negatively affected the flowability and setting time [18,36,37].

Figure 5.9 depicts the influence of Na_2O content on the setting time of GPMs. The GPMs activated at low NaOH molarity took significantly longer time to set. This is due to the slow rate of chemical reaction at low ambient temperature and little

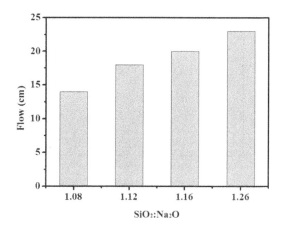

FIGURE 5.8 Impact of SiO_2:Na_2O on the flow of GPMs.

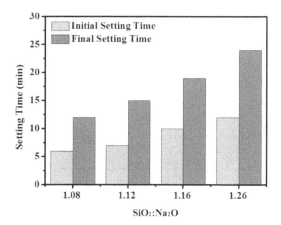

FIGURE 5.9 Impact of SiO_2:Na_2O on the setting time of GPMs.

amount of Na$_2$O content. In this study, the GPM mixtures that were prepared with high concentration of solution revealed very fast setting time. The GPMs setting time was further improved considerably with the reduction of NH molarity. Both initial and final setting times were enhanced with the decrease in NH molarity. Besides, the rate of setting was increased appreciably as indicated by the substantial difference in the initial setting time. The difference between initial and final setting time was also increased with the reduction of NH molarity in the mortar. This enhancement of setting time at lower NH molarity was attributed to the slower rate of setting [36].

Figure 5.10 presents the effect of S:B on the flow of GPMs. The flowability of mortar was increased with increasing S:B. An increase in the solution has increased the water content (H$_2$O:dry binder) and improved the workability (Table 5.3). Figure 5.11 represents the influence of solution content on the setting time of GPMs. As mentioned earlier, the GPM activated at high solution content was set at longer time

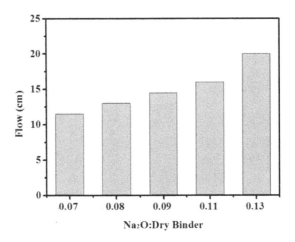

FIGURE 5.10 Impact of varying Na$_2$O:dry binder on the flow of GPMs.

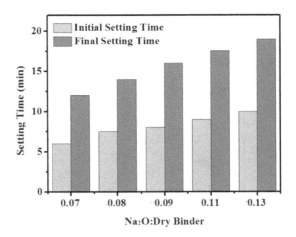

FIGURE 5.11 Impact of varying Na$_2$O:dry binder on the setting time of GPMs.

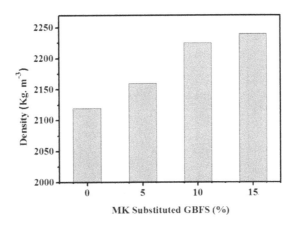

FIGURE 5.12 Effect of MK substituted by GBFS on GPMs density.

because of the slow rate of chemical reaction. The GPMs' setting time was improved considerably with the increase in solution content. Both initial and final setting time were also enhanced with the increase in S:B.

Figure 5.12 illustrates the effect of MK substituted by GBFS on the density of GPMs. The density of GMPs was found to increase with increasing percentage of MK replacement. The particle size and specific gravity of MK was found to influence the GPMs density. An increase in the content of Al_2O_3 and SiO_2 led to produce sodium aluminosilicate hydrate (N-A-S-H) gel beside the calcium–silicate–hydrate (C-S-H) which in turn improved the microstructure of GPMs as evidenced earlier [25].

5.5 COMPRESSIVE STRENGTH PERFORMANCE

Cubic moulds of size 50 mm, cylinder of dimension (75 mm × 150 mm) and prism of dimension (40 mm × 40 mm × 160 mm) were used to prepare compressive, tensile and flexural strength samples. Compressive, tensile and flexural strengths of GPMs were measured using ASTM C109/109M standard [30]; ASTM C496/C496M-11 [38] and ASTM C78 [39], respectively. The strengths were evaluated and compared with the control sample (cement mortar). The compressive strength of specimens was evaluated at the age of 1, 7 and 28 days. The averaging of three specimens is performed to present the results. Effect of MK replaced by GBFS and modulus of solution (SiO_2:Na_2O and Na_2O:dry binder) on the early compressive strength of GPMs were determined. The compressive strength of GPMs was measured after 1, 7 and 28 days using ASTM C109/C109M and averaged over three realizations.

Figure 5.13 shows the impact of MK substitution on the early compressive strength of GPMs. The compressive strength of MK replaced by GBFS sample after 1 day revealed lower values (27.6, 24.2 and 20.4 MPa) with 5%, 10% and 15%, respectively, than the one prepared without MK (32.8 MPa). However, after 28 days the MK-substituted samples (5, 10 and 15 mass%) achieved higher strength (62.5, 62.8 and 63.1 MPa) than the one prepared without MK (44.8 MPa). This observation was majorly ascribed to the increase in curing time and the completion

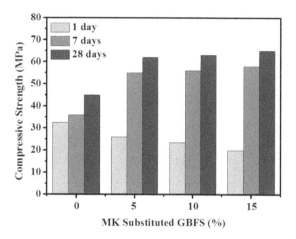

FIGURE 5.13 Impact of MK substituted by GBFS on the early compressive strength of GPMs.

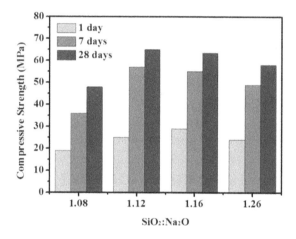

FIGURE 5.14 Impact of $SiO_2:Na_2O$ on the early compressive strength of GPMs.

of geopolymerization process. An increase in the content of Al_2O_3 and SiO_2 has improved the geopolymerization and produced N-A-S-H and aluminate-substituted calcium–silicate–hydrate (C-A-S-H) gels in addition to the C-S-H, thus enhanced the strength properties of GPMs [25,40]. GPM mixture with 5% MK replaced by GBFS showed a high early compressive strength compared with the other percentages, which was selected for the second stage.

Figure 5.14 displays the influence of varying NaOH molarity on the development of compressive strength. The effect of NH solution molarity on $SiO_2:Na_2O$ ratios when added with NS solution, the ratio of NS:NH and the direct effect of NH molarity on $SiO_2:Na_2O$ ratio are determined. An increase in the NH molarity has enhanced the Na_2O contents and reduced the silicate-to-sodium ratio as listed in Table 5.2. The compressive strength was increased with the $SiO_2:Na_2O$ ratio. Consequently, the compressive strength was related to the amount of NaOH in the alkaline solution.

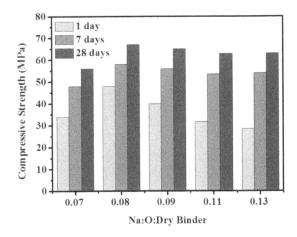

FIGURE 5.15 Impact of Na₂O:dry binder on the early compressive strength of GPMs.

As the SiO$_2$:Na$_2$O ratio was increased, the degree of dissolution and hydrolysis were accelerated, thereby inhibited the polycondensation. A solution molarity of 14 M (SiO$_2$:Na$_2$O of 1.16) showed high early strength compared to other molarities. Zuhua et al. [41] reported that the presence of high concentration of NH has accelerated the dissolution of silica and alumina and thus hindered the polycondensation. Samples prepared with 18 M of NH (low SiO$_2$:Na$_2$O of 1.08) displayed lower strength than those prepared with 10, 14 and 16 M [37]. Phoongernkham et al. [24] reported that the dissolution of calcium was suppressed at high NaOH concentration resulting in less hydration products. In addition, an excess hydroxide ion caused aluminosilicate gel precipitation at the early stage and resulted in lower strength GPs [42]. GPM mixture prepared with 14 M (SiO$_2$:Na$_2$O = 1.16) revealed higher early strength after 24 hours and thus selected for the next stage.

Figure 5.15 depicts the effect of Na$_2$O:dry binder on the development of compressive strength. An increase in the S:B content leads to an increase in the ratio of Na$_2$O:dry binder and H$_2$O:dry binder (Table 5.3). A reduction in the S:B was found to increase the early strength of GPMs. Samples prepared with 8% of Na$_2$O:dry binder and 23% of H$_2$O:dry binder exhibited the highest early strength of 47 MPa/24 hours and 63 MPa/28 days, respectively, compared to other ratios. In other words, the presence of too much water reduced the geopolymerization rate as reported by Zuhua et al. [41]. Huseien et al. [25] reported that an increase in the alkaline solution to binder could increase the water content and reduce the amount of C-A-S-H and N-A-S-H gels. Consequently, a poor structure could be produced with low early strength.

5.6 SPLITTING TENSILE STRENGTH

Figure 5.16 compares the splitting tensile strength of prepared GPMs with OPC mortar. Twelve cylindrical samples were prepared using the optimum results from Phase 3, where 5 wt.% of MK replaced by GBFS, 14 M of NH (SiO$_2$:Na$_2$O = 1.16) and 8% of Na$_2$O:dry binder were used. The tensile strength of all samples cured at ambient

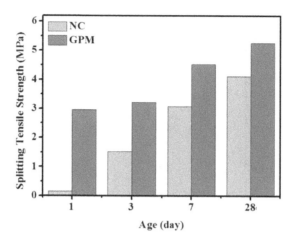

FIGURE 5.16 Curing time-dependent development of early split tensile strength of GPMs as compared to OPC mortar.

temperature revealed an increase with increasing curing time. Moreover, all the samples demonstrated higher early split tensile strength compared with control samples (OPC). After 24 hours, the GPM exhibited a split tensile strength of 2.95 MPa which was almost 10 times greater than that of OPC mortar (0.32 MPa). The ability of using GPM as a new alternative repair material was clearly depicted in the results of tensile strength. Similar trends were reported elsewhere [23].

5.7 FLEXURAL STRENGTH

Figure 5.17 presents the flexural strength of GPM prisms. Flexural strength was evaluated after 1, 3, 7 and 28 days and compared with OPC (as control samples). After 24 hours, GPM achieved a very high early flexural strength (5.7 MPa) in comparison

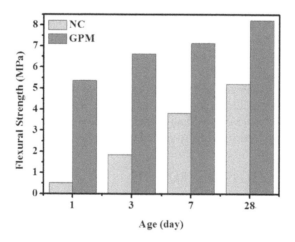

FIGURE 5.17 Flexural strength of GPMs as compared to OPC mortar.

to the control samples (OPC) which was as low as 0.6 MPa. The early flexural strength of GPM was further increased to 8.2 MPa after 28 days. The geopolymerization has contributed more SiO_2 and Al_2O_3 to the dissolution and produced the N-A-S-H and C-A-S-H gels besides the C-S-H gel. This explains the occurrence of higher strength of GPM compared to OPC sample which depended only on the C-S-H.

5.8 BOND STRENGTH PERFORMANCE

ASTM C882 [43] was depended to evaluate the shear bond strength capacity between the Portland cement substrate (NC) and GPM with stiffer slant shear angle 30°. For the casting of the specimens, the NC was casted and cured for 3 days in the water. Afterwards, these specimens were left in the laboratory (27°C and 75% relative humidity) till they were reached at the age of 28 days. Next, they are fixed in cylinder moulds (100 mm × 200 mm), casted for the second part (OPC and GPM) and then evaluated after 1, 3, 7 and 28 days. The shear bond strength was defined as the ratio of maximum load at failure and the bond area. The reported results of shear bond strength were considered as the average of three samples. The procedure of shear bond test was presented in Figure 5.18.

The bond strength between OPC (NC) and GPM was determined using a slant shear bond test. Cylinder slant shear specimens of dimension (100 mm × 200 mm) with interface line at 30° are prepared. The bond strength was tested at 1, 3, 7 and 28 days after curing at ambient temperature. The slant shear test is the most widely accepted test for the bonding of repair materials to concrete. The results of GPM bond strength are compared to OPC mortar (Figure 5.19). The bond strength of GPM that was prepared with 5 mass% of MK, 1.16 of $SiO_2:Na_2O$ and 8% of Na_2O:dry binder displayed the highest bond strength of 9.9 and 22.4 MPa at the early (1 day) and late age (28 days), respectively, when compared to OPC. Figure 5.20 illustrates the typical bond failure of a slant shear sample, where the bonding surface was found

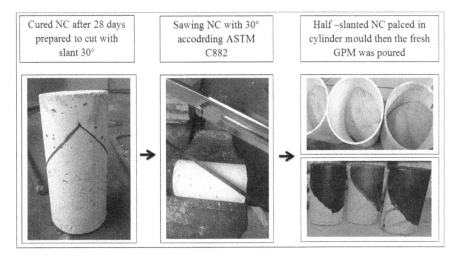

FIGURE 5.18 Preparation of composite cylinder (GPMs-NC) for bond strength test.

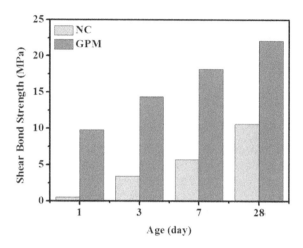

FIGURE 5.19 Shear bond strength between OPC (NC) and GPMs with interface line at 45° to the vertical.

FIGURE 5.20 Typical failure modes of GPMs.

to be still intact. The cracks were passed through the NC substrate and GPM interface. Moreover, there was no significant gap between the two bonding surfaces as confirmed by other report [23].

5.9 SUMMARY

The performance criteria of GPMs as concrete repair materials were evaluated. The impact of MK substituted by GBFS on the early mechanical properties of GPMs was determined to examine its feasibility for repair applications. GPM specimens were activated with varying solution content (Na_2O:dry binder and H_2O:dry binder) and alkaline solution modulus (SiO_2:Na_2O). Performance evaluation of GPMs was conducted at ambient temperature. Based on the achieved results, the following conclusions are drawn:

i. MK replaced by GBFS developed the workability of GPMs, where the setting time was increased and the density was reduced with increasing calcium and silicate contents.
ii. A reduction in the value of Na_2O:dry binder from 13% to 8% has allowed to develop the early compressive strength as much as 47.84 MPa/24 hours as the water content is reduced from 39% to 23%.
iii. Activation with 1.16 of SiO_2:Na_2O ratio of the alkaline solution achieved the highest early strength after 24 hours.
iv. These achieved GMPs have demonstrated higher compressive, tensile and flexural strength than that of OPC mortar.
v. The attainment of high bond strength of such GPMs indicated their ability as an alternative potential repair material.

REFERENCES

1. Green, M.F., et al., Effect of freeze-thaw cycles on the bond durability between fibre reinforced polymer plate reinforcement and concrete. *Canadian Journal of Civil Engineering*, 2000. **27**(5): pp. 949–959.
2. Lee, M.-G., Y.-C. Wang, and C.-T. Chiu, A preliminary study of reactive powder concrete as a new repair material. *Construction and Building Materials*, 2007. **21**(1): pp. 182–189.
3. Yang, Q., et al., Properties and applications of magnesia–phosphate cement mortar for rapid repair of concrete. *Cement and Concrete Research*, 2000. **30**(11): pp. 1807–1813.
4. Huseien, G.F., et al., Synthesis and characterization of self-healing mortar with modified strength. *Jurnal Teknologi*, 2015. **76**(1): pp. 195–200.
5. Al-Zahrani, M., et al., Mechanical properties and durability characteristics of polymer- and cement-based repair materials. *Cement and Concrete Composites*, 2003. **25**(4): pp. 527–537.
6. Mirza, J., et al., Preferred test methods to select suitable surface repair materials in severe climates. *Construction and Building Materials*, 2014. **50**: pp. 692–698.
7. Aleem, M. and P. Arumairaj, Geopolymer concrete: a review. *International Journal of Engineering Sciences & Emerging Technologies*, 2012. **1**(2): pp. 118–22.
8. He, Y., et al., Preparation of self-supporting NaA zeolite membranes using geopolymers. *Journal of Membrane Science*, 2013. **447**: pp. 66–72.
9. Wang, K., et al., Effects of the metakaolin-based geopolymer on high-temperature performances of geopolymer/PVC composite materials. *Applied Clay Science*, 2015. **114**: pp. 586–592.
10. Zhang, J., et al., Synthesis of a self-supporting faujasite zeolite membrane using geopolymer gel for separation of alcohol/water mixture. *Materials Letters*, 2014. **116**: pp. 167–170.
11. Zhang, Z., et al., Preparation and characterization of a reflective and heat insulative coating based on geopolymers. *Energy and Buildings*, 2015. **87**: pp. 220–225.
12. Davidovits, J., *Geopolymer chemistry and applications*. 2008: Geopolymer Institute, ISBN 978-2-9514820-1-2.
13. Zhang, Z., et al., Geopolymer from kaolin in China: an overview. *Applied Clay Science*, 2016. **119**: pp. 31–41.
14. Duxson, P., et al., The role of inorganic polymer technology in the development of 'green concrete'. *Cement and Concrete Research*, 2007. **37**(12): pp. 1590–1597.
15. Malhotra, V.M. and P.K. Mehta, *Pozzolanic and cementitious materials*. Vol. 1. 1996: Taylor & Francis.

16. Sabir, B., S. Wild, and J. Bai, Metakaolin and calcined clays as pozzolans for concrete: a review. *Cement and Concrete Composites*, 2001. **23**(6): pp. 441–454.
17. Van Deventer, J.S., J.L. Provis, and P. Duxson, Technical and commercial progress in the adoption of geopolymer cement. *Minerals Engineering*, 2012. **29**: pp. 89–104.
18. Bernal, S.A., et al., Evolution of binder structure in sodium silicate-activated slag-metakaolin blends. *Cement and Concrete Composites*, 2011. **33**(1): pp. 46–54.
19. De Silva, P. and K. Sagoe-Crenstil, The effect of Al_2O_3 and SiO_2 on setting and hardening of $Na_2O-Al_2O_3-SiO_2-H_2O$ geopolymer systems. *Journal of the Australian Ceramic Society*, 2008. **44**(1): pp. 39–46.
20. Chindaprasirt, P., et al., Effect of SiO_2 and Al_2O_3 on the setting and hardening of high calcium fly ash-based geopolymer systems. *Journal of Materials Science*, 2012. **47**(12): pp. 4876–4883.
21. Hu, S., et al., Bonding and abrasion resistance of geopolymeric repair material made with steel slag. *Cement and Concrete Composites*, 2008. **30**(3): pp. 239–244.
22. Pacheco-Torgal, F., J. Castro-Gomes, and S. Jalali, Adhesion characterization of tungsten mine waste geopolymeric binder. Influence of OPC concrete substrate surface treatment. *Construction and Building Materials*, 2008. **22**(3): pp. 154–161.
23. Phoo-ngernkham, T., et al., Effects of sodium hydroxide and sodium silicate solutions on compressive and shear bond strengths of FA–GBFS geopolymer. *Construction and Building Materials*, 2015. **91**: pp. 1–8.
24. Phoo-ngernkham, T., et al., High calcium fly ash geopolymer mortar containing Portland cement for use as repair material. *Construction and Building Materials*, 2015. **98**: pp. 482–488.
25. Huseien, G.F., et al., Influence of different curing temperatures and alkali activators on properties of GBFS geopolymer mortars containing fly ash and palm-oil fuel ash. *Construction and Building Materials*, 2016. **125**: pp. 1229–1240.
26. Huseien, G.F., et al., Effect of metakaolin replaced granulated blast furnace slag on fresh and early strength properties of geopolymer mortar. *Ain Shams Engineering Journal*, 2018. **9**(4): pp. 1557–1566.
27. Ismail, M., et al. Early strength characteristics of palm oil fuel ash and metakaolin blended geopolymer mortar. in *Advanced materials research*. 2013: Trans Tech Publ. **693**: pp. 1045–1048.
28. Astm, C., *Standard specification for fly ash and raw or calcined natural pozzolan for use as a mineral admixture in Portland cement concrete*. 2003. ASTM C, Philadelphia: 618–685.
29. Commitee, A., *C09. ASTM C33-03, Standard Spesification for Concrete Agregates*. 2003: ASTM International.
30. Standard, A., *A109, 2012, "Standard Test Method for Compressive Strength of Hydraulic Cement Mortars (Using 2-in. Cube Specimens,"* 2012: ASTM International, DOI 10.1520/A0307-12, <www. astm. org.Journal>
31. Sofi, M., et al., Engineering properties of inorganic polymer concretes (IPCs). *Cement and Concrete Research*, 2007. **37**(2): pp. 251–257.
32. Al-Majidi, M.H., et al., Development of geopolymer mortar under ambient temperature for in situ applications. *Construction and Building Materials*, 2016. **120**: pp. 198–211.
33. Lee, N., E. Kim, and H. Lee, Mechanical properties and setting characteristics of geopolymer mortar using styrene-butadiene (SB) latex. *Construction and Building Materials*, 2016. **113**: pp. 264–272.
34. Kumar, S., R. Kumar, and S. Mehrotra, Influence of granulated blast furnace slag on the reaction, structure and properties of fly ash based geopolymer. *Journal of Materials Science*, 2010. **45**(3): pp. 607–615.
35. Sugama, T., L. Brothers, and T. Van de Putte, Acid-resistant cements for geothermal wells: sodium silicate activated slag/fly ash blends. *Advances in Cement Research*, 2005. **17**(2): pp. 65–75.

36. Rattanasak, U. and P. Chindaprasirt, Influence of NaOH solution on the synthesis of fly ash geopolymer. *Minerals Engineering*, 2009. **22**(12): pp. 1073–1078.
37. Lin, K.L., et al., Effects of SiO_2/Na_2O molar ratio on properties of TFT-LCD waste glass-metakaolin-based geopolymers. *Environmental Progress & Sustainable Energy*, 2014. **33**(1): pp. 205–212.
38. Norma, A., *C496/C496M-11, Standard test method for splitting tensile strength of cylindrical concrete specimens*. 2012. Philadelphia, ASTM C: pp. 1–12.
39. ASTM, C., *Standard test method for flexural strength of concrete (using simple beam with third-point loading)*. 1999: American Society for Testing and Materials.
40. Dombrowski, K., A. Buchwald, and M. Weil, The influence of calcium content on the structure and thermal performance of fly ash based geopolymers. *Journal of Materials Science*, 2007. **42**(9): pp. 3033–3043.
41. Zuhua, Z., et al., Role of water in the synthesis of calcined kaolin-based geopolymer. *Applied Clay Science*, 2009. **43**(2): pp. 218–223.
42. Lee, W. and J. Van Deventer, The effects of inorganic salt contamination on the strength and durability of geopolymers. *Colloids and Surfaces A: Physicochemical and Engineering Aspects*, 2002. **211**(2): pp. 115–126.
43. ASTM, C., *882–91–Standard test method for bond strength of epoxy resin system used with concrete by slant shear*. 1991: American Society for Testing Materials, ASTM.

6 Compatibility of Geopolymer for Concrete Surface Repair

6.1 INTRODUCTION

Over the years worldwide, Ordinary Portland Cement (OPC) has been extensively used to bind concrete that is effective for various construction purposes. Several studies [1–3] reported that the concrete showed low durability to aggressive environments and led to much deterioration during life service. Wang et al. [4] reported that the cost of restorations and rehabilitations was close to or even exceeded the cost of new construction. The surfaces of concrete structures, such as sidewalks, pavements, parking decks, bridges, runways, canals, dykes, dams, and spillways, deteriorate progressively due to a variety of physical, chemical, thermal and biological processes. Actually, the performance of concrete compositions is greatly affected by the improper usage of substances, and physical and chemical conditions of the environment [2,5]. The immediate consequence is the anticipated need of maintenance and execution of repairs [6,7]. To overcome these issues and minimize these problems, researchers made dedicated efforts to develop many different types of repair materials, with or without OPC, such as emulsified epoxy mortars, sand epoxy mortars and polymer-modified cement-based mortars, to attain repair materials for damaged concrete structures. However, variation between the results obtained by different researchers could be attributed to the difference in raw materials, specimen geometry and test methods. For the construction purpose, the geopolymer mortar (GPM) is a newly introduced binder with much higher resistance to severe climatic conditions [8,9]. Currently, intensive research studies have been carried out to develop these binders with emergent high performance as sustainable construction materials [10–13].

Recently, there has been an active development in the areas of geopolymer pastes, mortars and concrete, which produce much lower carbon dioxide than using OPC and are therefore much better for the environment. Rather than the high levels of CO_2 created in the production of OPC, the alkaline method of making concrete using GPs [14–16] produces binders using wastes, such as palm oil fuel ash (POFA), metakaolin, fly ash (FA), ceramic tile wastes, glass wastes and ground blast furnace slag (GBFS). In recent years, this development in creating an alternative to OPC has caught the attention of professionals in the commercial and academic sectors [17,18]. The discussions have not only focused on the relatively similar functionality to OPC but also on the natural features of materials, such as its excellent strength performance, high resistance to sulphuric acid and sulphate attack, its ability to withstand heat, its reduced energy consumption and level of CO_2 emissions [19–22]. Some

DOI: 10.1201/9781003173618-6

researchers have centred on several alumina-silicate materials and their responses as well as microstructural classification when activated with various alkaline activators [23–26]. Research in the production of ecological pastes/mortars has stepped up in recent times [21,27], but there have not been any studies to develop an outline for ecological mortar mix proportioning for repair purposes.

Recent studies have indicated that the presence of calcium content in FA affects significantly the resultant hardening characteristic compressive strength (CS) of alkali-activated mortars (AAMs) [28–30]. Calcium oxide (CaO) generates calcium–silicate–hydrate (C-S-H) along with the sodium–aluminium–silicate gel (N-A-S-H) [31]. The compatible nature of aluminate-substituted calcium–silicate–hydrate (C-(A)-S-H) and N-A-S-H gels has considerable influence on the hybrid OPC and alkaline solution–activated aluminosilicate (AS) systems, wherein both products may be generated [32,33]. Earlier research studies have used the synthetic gels to determine the influence of high pH levels on each gel component. The aqueous aluminate was found to affect greatly the C-S-H product formation [34,35]. Besides, the aqueous Ca could modify the N-A-S-H gels and partially replace the sodium (Na) with Ca to produce (N, C)-A-S-H gels [34,36]. However, the mechanisms for the formation of such gels and subsequent improvements have not yet been completely understood. Furthermore, to explore the feasibility of achieving green cements for the construction purposes, both gels must co-exist and systematic studies on the compatible nature of N-A-S-H and C-A-S-H gels became essential. Several investigations have been carried out on the materials containing calcium compounds, especially ground-granulated blast furnace slag [37,38]. However, most of the earlier studies used high volume of corrosive sodium silicate (NS) and/or sodium hydroxide (NH) to produce AAMs, which posed health and safety issues to workers during handling these materials [39]. Other researchers [40] proposed a simple approach to produce environmentally friendly alkali-activated mortar with improved mechanical properties by overcoming the thermally activated processes and promoting an easy management.

The term compatibility has become a popular buzzword in the repair industry, as it implies durability of repairs in general and adequate load-carrying capacity in the case of concrete repairs [41]. It is, however, more than this. Compatibility can be defined as a balance of physical, chemical and electrochemical properties and dimensions between a repair material and the existing damaged substrate that will ensure that the repair can withstand all the stresses induced by volume changes and chemical and electrochemical effects without distress and deterioration over a designated period. In concrete repair work, the bond compatibility between repair materials and concrete substrate is one of the important factors affecting durability and sustainability of repairing work [42,43]. The coefficient of thermal expansion (CTE) is a measure of length change in a material when it is subjected to a variation in temperature. When two materials (repair material and substrate) of different coefficients of thermal expansion are joined together and subjected to significant temperature changes, stresses are generated in the composite material. Our literature study revealed that no studies have so far evaluated the mechanical properties of such AAMs in detail, specially and specifically to repair the damaged concrete surfaces and assess the compatibility between alkali-activated mortar as repair materials and deteriorated concrete substrate. Most of the studies have only analysed the mineralogy and microstructure properties of AAMs.

Driven by this idea, this chapter examines the feasibility of achieving a new type of high-performance GPM with improved mechanical and durable properties to repair damaged concrete surfaces. These new high-performance GPMs were designed using industrial wastes such as FA and GBFS in appropriate proportions. The effects of different ratios of FA replaced by GBFS and activated with low molarity of alkaline activator solution were investigated to determine the flowability and setting time, compressive and bond strength, and thermal compatibility with base concrete of the synthesized GPMs. Diverse tests including CS, tensile strength (TS) and flexural strength (FS) were performed to characterize the prepared GPMs, which could be advantageous to repair damaged concrete surfaces. Furthermore, several tests such as slant shear (SS) bond strength, bending stress, thermal expansion coefficient and third-point load-flexural measurement were also performed to evaluate the bond strength and compatibility between GPMs and the mortar/concrete substrates.

6.2 GEOPOLYMER PREPARATION

Pure GBFS was used (without any further purification) as a constituent to make the binder free of cement. This slag (off-white in appearance) was distinct from other auxiliary cementitious substances because of having both cementitious and pozzolanic traits. GBFS emerges from hydraulic chemical reaction when water is mixed. X-ray fluorescence spectra revealed that the GBFS is composed of calcium (51.8%), silicate (30.8%) and alumina (10.9%). FA (alumina-silicate substance) with low level of calcium was used to make GPMs. It satisfied the ASTM C618 requirements for FA Class F and appeared grey in colour with 5.2% of calcium, 57.2% of silicate and 28.8% of aluminium contents. The median of particle for FA and GBFS (obtained using particle size analyser) was 10 and 12.8 µm, respectively. The Brunauer–Emmett–Teller (BET) surface area of FA and GBFS was calculated to analyse the CS of their physical characteristics. The specific surface area of FA displayed the highest value of 18.1 m^2/g compared to GBFS (13.6 m^2/g).

Figure 6.1 shows the X-ray diffraction (XRD) patterns of FA and GBFS. FA revealed pronounced diffraction peaks around 2θ = 16°C–30°C, which were allocated to the presence of crystalline silica and alumina compounds. Nonetheless, the occurrences of other sharp peaks were assigned to the presence of crystalline phases of quartz and mullite. The absence of any sharp peak in the XRD pattern of GBFS indeed confirmed their true disorder. The occurrences of silica and calcium in the pattern were the significant factors towards GBFS creation. The presence of great amount of reacting amorphous Si and Ca in GBFS was advantageous for the synthesis of GPMs. However, inclusion of FA was needed to surmount the low level of Al_2O_3 (10.49 wt.%) in GBFS. Scanning electron microscope images of FA and GBFS exhibited that FA consisted of spherical particles with smooth surface and GBFS comprised irregular and angular particles, similar to the ones reported earlier [44].

Saturated natural sand from river (siliceous) was utilized as fine aggregates to prepare the proposed mortars. It was first cleaned using water to lessen the presence of silts and impurities as per ASTM C117 standard [45]. This was followed by oven

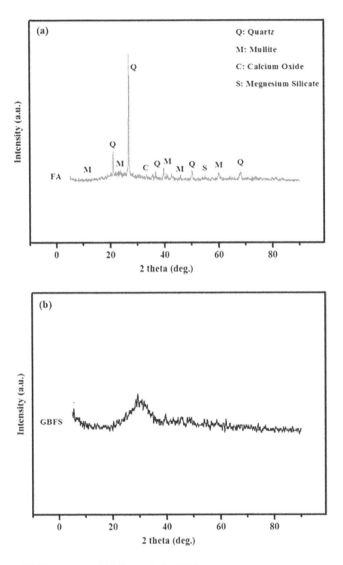

FIGURE 6.1 XRD patterns of (a) FA and (b) GBFS.

dried at 60°C for 24 hours to remove the moisture before being graded to match with the specifications of ASTM C33-33M [46]. The fineness modulus and specific gravity of the prepared aggregates were 2.9 and 2.6, respectively. Analytical-grade NH (98% purity) in the form of pellets was dissolved in water to prepare its solution with the concentration of 2M (7.4 Na_2O and 92.6% H_2O). Analytical-grade NS blend made of SiO_2 (29.5 wt. %), Na_2O (14.70 wt.%) and H_2O (55.80 wt.%) was utilized. The prepared 4M NH solution was kept for 24 hours to cool down. It was then mixed with NS solution to achieve the ultimate alkaline mixture having $SiO_2:Na_2O$ ratio of 1.2 wherein the proportion of NS:NH for all alkaline mixtures was maintained at 0.75. In this study, the total Na_2O, SiO_2 and H_2O were 10.53, 12.64 and 76.8 (by weight,

TABLE 6.1
Mix Design of Proposed GPMs (mass%)

		Mix Design Formulation of Alkali-Activated Mortars				
Materials (mass%)		$GPMs_1$	$GPMs_2$	$GPMs_3$	$GPMs_4$	$GPMs_5$
Binder (B)	FA	70	60	50	40	30
	GBFS	30	40	50	60	70
	$SiO_2:Al_2O_3$	2.10	2.15	2.22	2.29	2.38
	$CaO:SiO_2$	0.38	0.51	0.65	0.80	0.98
B:A		1.0	1.0	1.0	1.0	1.0
S:B		0.40	0.40	0.40	0.40	0.40
Na_2SiO_3:NaOH		0.75	0.75	0.75	0.75	0.75
Sodium hydroxide (NaOH)	Molarity, M	2.0	2.0	2.0	2.0	2.0
	H_2O	92.6	92.6	92.6	92.6	92.6
	Na_2O	7.4	7.4	7.4	7.4	7.4
Sodium silicate (Na_2SiO_3)	H_2O	55.8	55.8	55.8	55.8	55.8
	Na_2O	14.7	14.7	14.7	14.7	14.7
	SiO_2	29.5	29.5	29.5	29.5	29.5
Total H_2O in alkaline solution		76.8	76.8	76.8	76.8	76.8
Solution modulus (Ms), $SiO_2:Na_2O$		1.2	1.2	1.2	1.2	1.2

S:B: alkaline solution-to-binder ratio, B:A: binder-to-fine aggregates ratio, NS:NH: sodium silicate-to-sodium hydroxide ratio.

%), respectively, compared to 20.75, 21.07 and 58.2 of solution prepared for 14 M of NaOH and 2.5 of Na_2SiO_3:NaOH ratio which are recommended in previous studies as the optimum ratios [47,48]. Regarding the content of Na_2O, SiO_2 and H_2O, the prepared alkaline solution is environmentally friendly, low cost, low energy consumption and low carbon dioxide emissions.

For all GPM mixtures, the values of alkaline solution-to-binder (S:B) and binder-to-fine aggregate (B:A) were selected to be 0.40 and 1.0, respectively, which were fixed for all mixes depending on trial mixtures as no standard exists for GPMs. Two types of industrial waste materials (FA and GBFS) were used to prepare GPM mix design. FA and GBFS were blended to determine the influence of CaO on the geopolymerization process. The GBFS content was kept between 30% and 70% as a source of CaO. In this study, blend containing high-volume FA such as 100% not considered as the mixture required high molarity of alkaline solution (14 M) and high temperature for curing (70°C). Likewise, the content of GBFS kept to 70% or less as the high volume of GBFS effects on the setting time and reduces it below 5 minutes. So, blend containing 70% of FA was prepared as control sample (Table 6.1). The NH molarity, the NS:NH and the modulus of alkaline solution $SiO_2:Na_2O$ (Ms) were kept constant for all mixes. The effects of varying levels of GBFS as a replacement of FA on the contents of SiO_2, CaO, Al_2O_3, $SiO_2:Al_2O_3$ and $CaO:SiO_2$ in AAMs are depicted in Table 6.1. Five substitution levels were implemented to assess the impact of CaO on the geopolymerization process. The CaO content improved with the increase in GBFS. Furthermore, CaO

TABLE 6.2
Strength Properties of Normal Concrtet Substrate

Engineering Properties at the Age of 28 Days	
Compressive strength (MPa)	43.6
Splitting tensile strength (MPa)	4.4
Flexural strength (MPa)	5.5
Porosity (%)	10.2
Dry shrinkage (micro-strain)	340

content augmented from 19.2% to 37.8% with the replacement of FA by GBFS from 30% to 70%, respectively. Conversely, both the SiO_2 and Al_2O_3 contents reduced with the increase in GBFS level depending on the chemical composition.

Following the ASTM C109/C109M-16a specification [49], cube moulds of dimension (50 mm × 50 mm × 50 mm) were prepared for the CS test. Prism specimens of dimension (40 mm × 40 mm × 160 mm) were prepared for the FS test. For the TS and the SS tests, cylindrical samples (diameter = 75 mm and depth = 150 mm) were casted. For the shrinkage test, prism specimens of dimension (25 mm × 25 mm × 250 mm) were prepared. Prior to casting, engine oil was applied on the inner surfaces of moulds to make the de-moulding job easier. A homogeneous mixture of NH and NS (by weight) was prepared, followed by a cool down at room temperature. Next, the AAMs (homogeneous mix of fine aggregates) were synthesized by mixing (via mortar mixer) FA and GBFS for 2 minutes under dry state. The obtained mix was further activated by incorporating the alkali solution. The entire matrix was blended for another 4 minutes using machine operated at an average speed. The prepared fresh mortar was cast in moulds in two layers, where each layer was consolidated using vibration table for 15 seconds to eliminate air voids [50]. Keeping in mind the Singapore weather, the AAM specimens were then left at a temperature of 27°C ± 1.5°C and relative humidity of 75%, for a period of 24 hours, to allow them to cure. Then, the specimens were taken out from moulds and left in same condition till the testing time. The OPC was procured from Holcim Cement Manufacturing Company complying to specified ASTM C150 standard. The OPC was utilized to prepare high-strength concrete substrates (≥40 MPa) to show the ability of using alkali-activated mortar as high-performance repair materials, where the trial mixture of cement:sand:gravel ratio was 1:1.5:3.0 and water-to-cement ratio (W:C) was 0.48. Table 6.2 depicts the strength properties of normal concrete (NC) substrate.

6.3 WORKABILITY OF FRESH GPMS

The workability of mortars that included flow diameter, initial and final setting time was determined by flow table procedure with modification from ASTM C230/C230M-14 [51] and ASTM C191 [52], respectively. Figure 6.2 shows the mortar flow with the addition of GBFS into the binder. All the five mixtures were stimulated with the same amount of geopolymer mixture. Results of mortar flow showed that increasing content of FA could reduce the workability (flowability) of GPMs. The

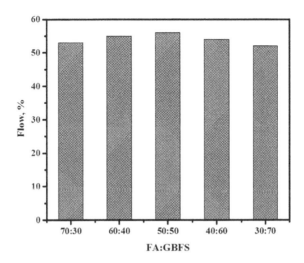

FIGURE 6.2 FA:GBFS ratio–dependent flow of GPMs.

mix containing GBFS and FA in 50:50 ratios exhibited the maximum flow percentage of 56. Furthermore, an increase in the FA level from 50% to 70% was found to reduce the flowability of the mortars from 56% to 52%, respectively. Generally, the workability of mortar is influenced by the specific surface area of the binder. The specific surface area of FA ($18.2\,m^2/g$) was higher than that of GBFS ($13.6\,m^2/g$). However, high water adsorption by FA having porous structure could lead to the workability drop of AAMs [53]. This effect was more pronounced for GBFS content over 50%, which can be attributed to the distinctions in the physical and chemical behaviour of the mixes. When GBFS amount was increased, the density of angular particles was enhanced and spherical ones were decreased. Moreover, the rapid rate of chemical reaction of high-volume GBFS content effects to decrease the plasticity of mixture which reduces the workability of the mortar [50]. Thus, further enhancement of GBFS level from 50% to 70% correspondingly led to a decrease in flow from 56% to 52%. Three factors such as specific surface area, particles' morphology and the chemical reaction rate could primarily be ascribed to the positive effect towards the workability improvement of GPMs. Hence it is clear that incorporating of GBFS and FA in GPMs (50:50) led to reduce the water absorption and enhance the plasticity of mixtures.

Figure 6.3 shows the effects of FA:GBFS on the setting time of GPMs. Mortars containing higher GBFS level showed faster initial and final setting time than the ones with lower slag level. When slag was included into the mixes, the setting time of GPMs was elongated considerably where both initial and final setting times were shortened with increasing slag contents. The initial and final setting time of GPMs incorporating GBFS in the range of 30%–70% varied between 24 and 11.5 minutes and 36 and 21 minutes, respectively. Besides, the initial and final setting time was minor and decreased with the increase in GBFS content. Results displayed that for GPMs containing low level of GBFS binder, it has significantly long setting time due to slow rate of chemical reaction [54]. In fact, an increase in FA amount enhanced the

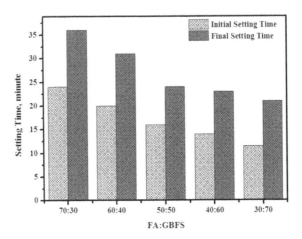

FIGURE 6.3 Effect of FA:GBFS ratio on the initial and final setting times of GPMs.

SiO_2 and Al_2O_3 contents and reduced the CaO content (Table 6.1) which negatively affected the rate of chemical reaction to produce C-S-H, C-A-S-H and N-A-S-H gels. This observation was also supported by the fact that higher slag content in the mortar could cause faster rate of setting [55,56].

6.4 STRENGTH PERFORMANCE

Generally, it is known that the repair materials must possess similar or better strength features than the substrate concrete. For repair mortar, ASTM C 109 standard practice was utilized to determine the CS. Samples were tested at the curing ages of 1, 3, 7, 28, 56 and 90 days. By combining the external load, volume alteration and characteristic CS mismatch between the repair substance and the substrate concrete, a tensile force can be produced in the repair material. When such forces generate a TS in addition to the tensile capacity of the repair component, failure of the material including tensile cracks, spalling or de-bonding is likely to occur. Therefore, TS is a vital attribute for the selection of an appropriate repair material. The TS of the substrate mortars and the repair component was evaluated using cylinders of dimension 75 mm × 150 mm according to the standard test method of ASTM C 496. The TS of the repair component was evaluated at the curing age of 1, 3, 7 and 28 days. Meanwhile, the TS of the concrete was assessed at the curing age of 28 days. Extra cylinders of the substrate mortar were examined for their TS alongside the SS tests performed on composite cylinders to determine the bond strength of the repair substance. FS also called as modulus of rupture, bend strength or fracture strength is a measure of mechanical attribute of brittle materials. This characterizes the material's ability to resist deformation under the applied load. The FS test was carried out using ASTM C78/C78M procedure, where an adequately cured (at ages of 1, 3, 7 and 28 days) prism specimen of dimension (40 mm × 40 mm × 160 mm) was used. Suitable repair materials must present an FS similar to or higher than the concrete substrate. The modulus of elasticity (MOE) test was performed on sufficiently cured

Compatibility of Geopolymer

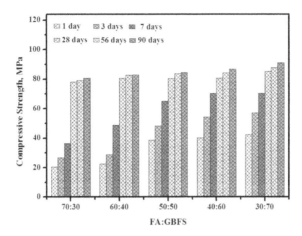

FIGURE 6.4 Effect of FA:GBFS ratios on the CS of GPMs.

(at age of 28 days) cylindrical specimens of dimension (100 mm × 200 mm) following the method documented in ASTM C469/C469M, 2010. Three sets of specimens were tested for each curing age and their average is reported. FS test is calculated using the following equation:

$$\sigma = 3FL/2bd^2 \tag{6.1}$$

where F is the load (force) at the fracture point (N), L is length of the support span (mm), b is the width (mm) and d is the thickness (mm).

Figure 6.4 illustrates the influence of GBFS content on the CS of prepared GPMs. The increasing GBFS contents showed positive effect on the CS. As the GBFS amount increased from 30% to 70% in GPMs, the early CS of specimens also increased to achieve 42.5 MPa at the age of 24 hours compared to 20.6 MPa achieved with 70% FA. Similar trends in the CS development were also observed at the curing ages of 3, 7, 28, 56 and 90 days. The CS values observed were 78.2, 80.4, 80.5, 80.7 and 85.1 MPa at the age of 28 days, and 80.6, 82.8, 84.5, 86.7 and 90.9 MPa at the age of 90 days for GBFS level of 30%, 40%, 50%, 60% and 70%, respectively. The gain in the CS of GBFS specimens was 47%, 64%, 77% and 94% of the final strength at 1, 3, 7 and 28 days. This enhancement in the CS with the increase in GBFS level was ascribed to the rise of CaO contents and reduction of SiO_2 levels in the mortar matrix. Furthermore, an enhancement in the GBFS content could generate high proportion of CaO to SiO_2 up to 0.95, leading to the higher amount of C-(A)-S-H gel formation in the mixtures containing up to 30% GBFS than the mixtures with above 30% addition of GBFS [48,50,57,58].

It is established that GBFS is an amorphous and granular substance. It consists of SiO_2, Al_2O_3, MgO and high CaO (51.8%) which allows it to form hydrated C-S-H gel as a main reaction product with 1.68 calcium-to-silicate ratio [32]. On incorporation of GBFS, the generated C-S-H gel further improves the CS of GPMs [32,59,60]. Kumar et al. [55] demonstrated that, at ambient temperature, the reaction could dominate by the dissolution and precipitation of C-S-H gel because of alkali activation of GBFS.

The achieved modification in the setting times and CS was majorly ascribed to the creation of cementitious C-S-H gel which in turn enhanced the hardened properties of AAMs. Influences of GBFS content on the CS development have been studied by Al-Majidi et al. [61]. The strength of GPM matrix was found to improve even at early ages due to the increase in GBFS content. Weiguo et al. [62] investigated the influence of GBFS content on the GPM matrix and reported an improvement in the CS and FS of the mixes with increasing GBFS content. It was acknowledged that the presence of GBFS could accelerate the hydration process and the formation of C-S-H gel [63]. Furthermore, an increased dosage of slag could strongly accelerate the hydration and enhance the mortars' strength.

The influence of GBFS on the gel formation was explained using three basic processes. The first process involved the enhanced CS due to an elevated generation rate of C-S-H gel originating from the addition of dissolved Ca existed on the GBFS surface. It was acknowledged that the higher rate of C-S-H formation in the existence of Ca could result in water shortage in the mortar matrix and raise its alkaline level, thus allowing elevated dissolution of existed alumina silicate [54,55,64]. The second mechanism could be related to the alkali-activated product of GBFS that was usually predominated by C-A-S-H gel. The subsistence of Al ions led to higher level of polymerization and notable cross-linking among C-S-H chains. In addition, the generation of N-A-S-H type of gel was considered as the third process to enhance the mortar's strength. Certainly, N-A-S-H being the trivial secondary product that coexisted in the composition domain of primary C-S-H gel category [65]. This could increase the gel compactness by reducing the overall porosity volume and thereby improving the CS of mortars [66].

Figure 6.5 shows the effect of $SiO_2:Al_2O_3$ and $CaO:SiO_2$ ratios on the CS of GPMs containing GBFS which was appreciably influenced by the increase in such ratios. The developed strength was directly correlated to the silicate-to-aluminium ratio, which was recorded to be optimum (42 MPa) for the silicate-to-aluminium ratio of 2.4 at the curing age of 24 hours. The strength of GPMs at the age of 28 days was the highest (85 MPa) for calcium-to-silicate ratio of 0.98. For silicate-to-aluminium

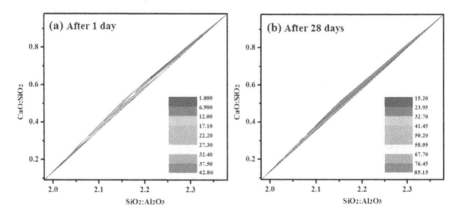

FIGURE 6.5 Influence of $SiO_2:Al_2O_3$ and $CaO:SiO_2$ ratios on the CS of GPMs at the ages of 1 and 28 days.

ratio of 2.22 and 2.29, the CS of GPM was reduced to 80.5 and 80.7 Mpa, respectively, at the age of 28 days. It was demonstrated that the CS of GPM reduced as the silicate level rose while the calcium and aluminium contents were reduced. It was reported [55,67] that an increase in CaO and Al_2O_3 contents from 22.7% and 6.3% to 39.3% and 8.9%, and SiO_2 decreased from 40.8% to 54.2% respectively, could indeed enhance C-(A)-S-H products and thus increase the CS of GPMs.

Figure 6.6 depicts the XRD pattern of GPMs with varied percentages of GBFS. The intensity of C-S-H peaks located at 28° and 31° was enhanced with the increase in GBFS content (from 0% to 70%). This led to the appearance of a new peak at 24° together with the replacement of gismondine ($CaAl_2Si_2O_8 \cdot 4(H_2O)$) peak by quartz phase. At 70% of GBFS, the mullite peak located at 16° showed less intensity compared to 60% GBFS samples, and nepheline peak positioned at 34°. Overall, an enhancement in the GBFS level led to the production of more dense gels (such as C-S-H and gismondine) and enhanced the microstructure and strength of GPMs.

Figure 6.7 displays the Fourier transform infrared spectroscopy (FTIR) spectra of GPMs synthesized with high volume of GBFS binder at different levels. Table 6.3 lists the FTIR band locations alongside band assignments. The FTIR spectra of GPMs obtained with varying contents of GBFS (as replacement to FA) revealed several bands corresponding to various functional groups. For 30% of GBFS, the bands located at 775.2, 873.6 and 989.5 cm^{-1} were shifted to 768.5, 871.8 and 956.9 cm^{-1}

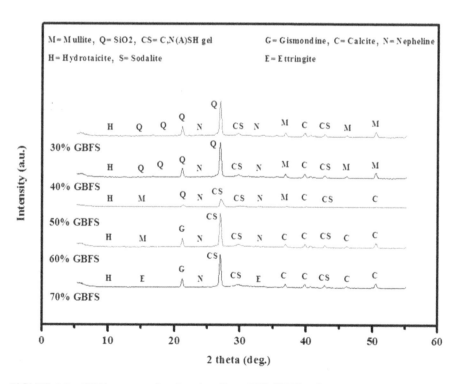

FIGURE 6.6 XRD patterns showing the effect of FA:GBFS ratios on the crystalline structures of GPMs.

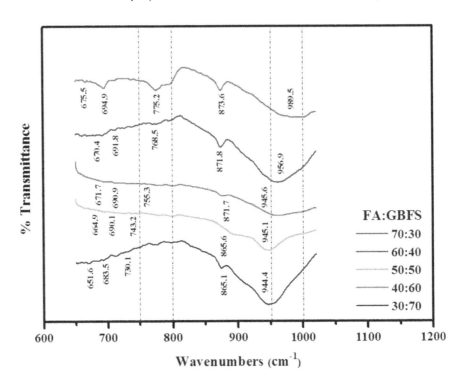

FIGURE 6.7 Effect of FA:GBFS on FTIR of GPMs.

TABLE 6.3
FTIR Band Locations and Allocations for GPM Containing GBFS

Mix				Band Positions (cm⁻¹) and Assignments				
FA:GBFS	Si:Al	Ca:Si	f_c (MPa)	Al-O	Si-O	AlO_4	Compressive Strength	C(N) ASH
70:30	2.10	0.39	78.2	675.5	694.9	775.2	873.6	989.5
60:40	2.15	0.51	80.5	672.4	691.8	768.5	871.8	956.9
50:50	2.22	0.65	80.46	671.7	690.9	755.3	871.7	945.6
40:60	2.29	0.80	80.68	664.9	690.1	743.2	865.6	945.1
30:70	2.38	0.97	85.09	651.6	683.5	730.1	865.1	944.4

(for 40% of GBFS); 755.3, 871.5 and 945.6 cm⁻¹ (for 50% of GBFS); 743.7, 865.6 and 945.1 cm⁻¹ (for 60% of GBFS) and 730.1, 865.1 and 944.4 cm⁻¹ (for 70% of GBFS). The decrease in the wavenumber of bands could be related to CS enhancement of AAMs from 80.5 to 85.1 MPa in the studied concentration range of GBFS (30%–70%). Moreover, the formation of AlO_4, C-(A)-S-H and N-A-S-H products displayed inverse proportionality with the wavenumbers. Low wavenumber implied the dissolution of more Si and the generation of AlO_4, C-(A)-S-H and N-A-S-H [68]. Similar

Compatibility of Geopolymer

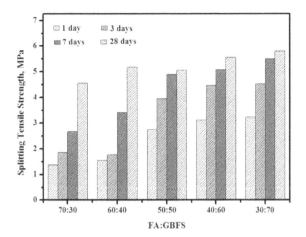

FIGURE 6.8 Effect of varying FA:GBFS ratio on splitting TS of GPMs.

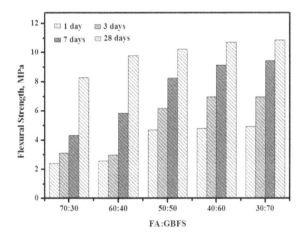

FIGURE 6.9 Effect of FA:GBFS ratios on the FS of GPMs.

trends were also observed at 50%, 60% and 70% of GBFS as replacement to FA, where an increase in the former diminished the band frequency.

Figure 6.8 illustrates the effect of GBFS content on the splitting TS of GPMs. Most of the results showed that an increase in GBFS content from 30% to 70% in the blended GPMs with FA indeed enhanced the splitting TS of specimens both at the early and later ages. The strength values recorded were 4.6, 5.1, 5.1, 5.6 and 5.8 MPa for 30%, 40%, 50%, 60% and 70% of GBFS, respectively. Furthermore, the increase in splitting TS showed the same trend as the CS of GPM. Similar results were also reported by Islam et al. [69]. Briefly, the incorporation of GBFS as a replacement of FA presented higher splitting TS than concrete substrate (4.4 MPa).

Figure 6.9 displays the FS of GPMs containing various percentages of GBFS. The respective FS values were determined for a curing period of 1, 3, 7 and 28 days. A minor change in the FS (from 4.7 to 4.9 MPa) was observed at 24 hours when GBFS

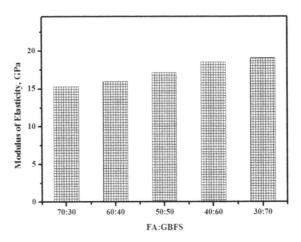

FIGURE 6.10 Effect of varying FA:GBFS ratio on the MOE of GPMs.

content increased from 50% to 70%. Similar trend in the strength development of GPMs was also observed at the age of 28 days. Here the specimen achieved a maximum strength of 10.9 MPa with increased GBFS content. The influence of high FA contents on the FS of GPMs was observed only at the early age. Similar trend was evidenced at the age of 28 days and the maximum strength recorded was 10.9 MPa for the highest GBFS content (70%). The higher volume of GBFS incorporation into the mix presented a continuous increase in the strength compared to the blend containing only FA. The observed enhancement in the FS of GPMs containing higher GBFS content could mainly be attributed to the increment in the CaO level in the mortar network.

Figure 6.10 depicts the FA:GBFS ratio–dependent variation (positive influence) in the MOE of AAMs. The MOE values of the studied AAMs were augmented from 15.2 to 19.1 GPa with the enhancement of GBFS level from 30% to 70%, respectively. It can be observed that the MOE of the GPMs increased together with the increase in CS, which is consistent with the other report [69]. Most of AAMs revealed MOE values lower than the concrete substrate (NC).

6.5 SLANT SHEAR BONDING STRENGTH

The bond strength between GPMs and concrete substrate was evaluated using the SS bonding test, where the hardened NC was diagonally slanted at 30° inclinations from the vertical. According to ASTM C882, the recommended bond angle of 30° represents the failure stress corresponding to a smooth surface closer to the minimum stress. The concrete substrates were prepared using the aged SS concrete prisms by cutting in half at 30° line to the vertical. The saw cut surface was used as it was shown to be a suitable substrate concrete surface for shear bond strength assessment. Half slanted NC was placed into the cylinder mould and fresh GPM was poured into the mould. This test was conducted using a compression machine on the specimens at the curing ages of 1, 3, 7 and 28 days. Figure 6.11 displays the step-by-step procedure of the SS test.

FIGURE 6.11 Preparation steps of composite GPM-NC SS specimens.

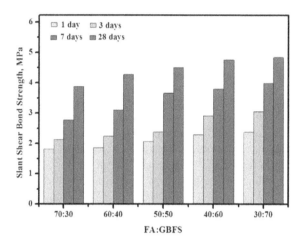

FIGURE 6.12 Influence of FA:GBFS ratios on the SSBS of GPMs at varied curing ages.

Figure 6.12 illustrates the FA:GBFS ratio–dependent changes of SS bond strength (SSBS) of GPMs at different curing ages. As can be seen from the results, the bond strength increased with the increase in curing age. At the early age (24 hours), the value of bond strength was increased from 1.8 to 2.3 MPa as the level of GBFS was rising from 30% to 70% in mortar matrix. At the age of 28 days, the 30° SS load-carrying capacity of NC substrate and GPM was observed in between 3.9 and 4.8 MPa when GBFS content was increased from 30% to 70%, correspondingly. The mix prepared with 70% of FA revealed much lower SSBS than those containing high levels of GBFS, indicating that the GPM containing the latter could produce higher amount of C-(A)-S-H gel than the one with lower GBFS content. This observation was attributed to the non-reactive nature and low content of silicate. Furthermore, the bond strength was enhanced with the increase in GBFS replacement to FA. The results of 30° bond strength for all GPMs displayed excellent outcome, wherein the failure zone occurred outside the bond zone [70].

6.6 THERMAL EXPANSION COEFFICIENT

The CTE of GPMs was measured from the alteration of length after subjecting them to temperature variation. After connecting the GPM and substrate composite of dissimilar CTE together and subjecting to appreciable temperature variation, stress was

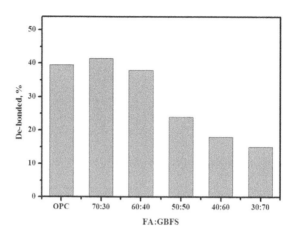

FIGURE 6.13 FA:GBFS ratio–dependent de-bonded percentage of studied GPMs.

produced in the composite. The resultant stress could cause a failure at the composite interface or within the material of lower strength. GPM must have a CTE very much alike the substrate concrete to circumvent the failure at elevated fluctuating temperature. The CTE is an important attribute of the repairing component subjected to varied temperature. The thermal consistency test of all the GPMs (acted as repairing material) with the base concrete was carried out by modifying ASTM C884 (intended for thermal compatibility between concrete and an epoxy resin overlay). The concrete block of dimension (80 mm × 100 mm × 200 mm) each served as a base for testing. These concrete slabs could sustain 10 freeze–thaw cycles needed by ASTM C666 test method (intended for concrete resistance evaluation to rapid freeze–thaw cycles called Method A). Various repairing mortars were implemented on the substrate concrete blocks with thickness ranging from 10 to 12 mm with a curing age of 28 days. These GPM-embedded concrete slabs were then subjected to 10 freeze–thaw cycles, wherein the temperature of each cycle varied from 5°C to −20°C. At the end of all cycles, each specimen was tested qualitatively for de-bonding together with visual inspection to trace any sign of crack, scaling or bond breakage between the base concrete and the repairing mortar. The percentage of non-bonded surface area was evaluated by lightly tapping the surface of repair material using small hammer. Empty sound was detected at locations where ever the repair product was de-bonded from the substrate surface. A rough estimate involving the de-bonding was made, wherein the percentage of de-bonded mortar after each cycle was estimated.

Figure 6.13 displays the FA:GBFS ratio–dependent de-bonded percentage of the studied GPMs. Mortars containing high volume of GBFS revealed good thermal compatibilities, which virtually remained unchanged even after 25 freeze–thaw cycles. Furthermore, it is demonstrated that the de-bonded percentage decreased and the compatibility enhanced between the GPMs and the concrete substrate with the increase in GBFS levels in the GPMs. Interestingly, the de-bonded percentage dropped from 40% to 12% when the GBFS level was increased from 30% to 70%, respectively, as revealed by the de-bonded patterns (Figure 6.14). On top of this, the

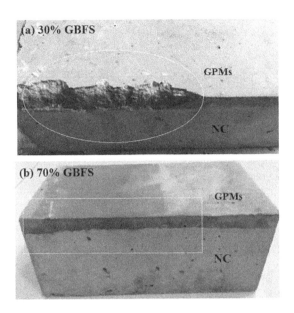

FIGURE 6.14 GBFS content–dependent de-bonded failure of the studied GPMs.

repair material (GPM) containing 40% of GBFS and above displayed an excellent resistance to thermal expansion and very high compatibility with the concrete substrate compared to cement mortar (OPC) repair agent.

6.7 THIRD-POINT LOADING FLEXURAL

This test was performed using two procedures to determine the compatibility between GPMs and concrete substrate. In the first procedure, the GPM was applied to a depression created at the bottom of a prism-shaped concrete (Figure 6.15a). Then, the specimen was tested identical to ASTM C 78 standard. The strength of the concrete substrate was 43.6 MPa at the curing age of 28 days. During the test, filled side of GPM (repair material) was positioned at the bottom of the specimen. For the second procedure, the vertical shear bonding strength between GPM and NC was tested (Figure 6.15b). The specimens were prepared by casting the prism concrete grade of 40 MPa (C40) having dimension (100 mm × 100 mm × 500 mm). Next, these samples were cut from the middle with the width of 30 mm before being placed in the prism mould to pour fresh GPMs. They were tested after the curing age of 28 days using third-point load at the loading rate of 0.2 kN/s. The compatibility or incompatibility of the repair materials with the concrete substrate was assessed by their failure mode. The GPM was declared compatible when the failure passed through the repair material and concrete substrate at the middle third of the beam. Otherwise, the GPM was approved as incompatible with the substrate concrete. The FS of the concrete substrate was analysed at the curing age of 56 days equivalent to 28 days of curing test of repair materials.

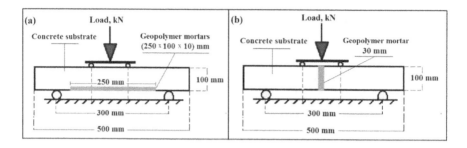

FIGURE 6.15 Diagram displaying the beam compatibility.

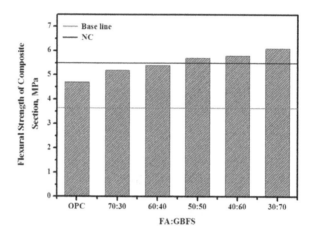

FIGURE 6.16 Binder-dependent FS of the composite section (GPMs-concrete substrate).

Figure 6.16 elucidates the FS of the composite beam (GPM-embedded concrete substrate) as a function of FA:GBFS ratios. It is known that rigid materials deflect lower in the flexural test than fragile one under similar load. In the present composite structure, the FS value was observed to increase with the increase in GBFS content. Composite beams containing 50% of GBFS and above achieved higher FS than the NC (5.6 MPa). The recorded FSs varied from 5.7 to 6.1 MPa with increasing GBFS level from 50% to 70% correspondingly. Nevertheless, reductions in the GBFS content below 50% have affected the beam resistance and lowered the strength of GPM below the one possessed by the concrete substrate beam. The GBFS level–dependent influence on the failure mode was also evaluated to determine the compatibility of these materials for repairing work (Table 6.4). Mortars prepared with high content of GBFS (60 and 70%) showed type A failure mode. A reduction in the GBFS content (50% and 60%) transformed the failure zone from A to B. The composite beam prepared with 70% of FA and cement mortar (OPC) displayed failure zone type C and D, respectively. Furthermore, the mortar prepared with 40% of GBFS and above disclosed high compatibility which is advantageous for repairing work.

Figure 6.17 displays the FA:GBFS ratio–dependent FS of the repairing beam (GPMs-concrete substrate) obtained using three-point load FS test. The GPM

TABLE 6.4
GBFS Content–Dependent Third-Point Loading Results and Failure Zone of the Composite Beam

	Flexural Strength (MPa)				
Mix	Repair Material	Base Line	Concrete Substrate	Composite Beam	Failure Zone
OPC	5.1	3.6	5.6	4.7	D
$GPMs_1$	8.2	3.6	5.6	5.2	C
$GPMs_2$	9.7	3.6	5.6	5.4	B
$GPMs_3$	10.2	3.6	5.6	5.7	B
$GPMs_4$	10.7	3.6	5.6	5.8	A
$GPMs_5$	10.8	3.6	5.6	6.1	A

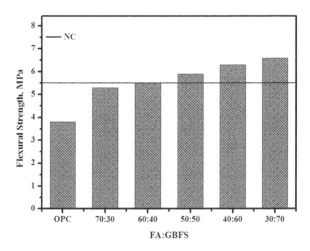

FIGURE 6.17 FA:GBFS ratio–dependent FS of the repairing beam (AAMs-concrete substrate).

containing 40% of GBFS and above revealed an excellent resistance and bond strength with NC than the one with 30% GBFS and cement mortar (control sample). The bond strength values were augmented with increasing GBFS level. An increase in the GBFS content from 30% to 70% improved the bond strength from 5.3 to 6.6 MPa, correspondingly as compared to the control sample (3.8 MPa). Besides, the failure zone was transformed from Type C to B and A with increasing GBFS content from 30%, 40% and 50%, and 60% and 70%, respectively compared to failure Type D which was achieved with ordinary cement mortar (OPC). In short, the failure zone was greatly influenced by the FA:GBFS ratio, where an increasing GBFS level could lead to an enhancement in bond strength and compatibility with the concrete substrate. The GPMs prepared with 30% and 40% of GBFS exhibited lower FS than that of NC (5.6 MPa).

6.8 BENDING STRESS

For this test, concrete specimens identical to the fracture test were prepared (Figure 6.18) following Phoongernkham et al [28] and casted by combining and filling GPMs within the notch, which were then left at room temperature for curing till the age of 28 days. Three-point bending test was conducted with deflection-adjusted load rate of 0.2 kN/s and analysed from the averages of three samples. Figure 6.19 illustrates the failure zone types in three-point loading and bending stress of the composite beam. The compatibility between GPMs and NC, the CTE, third-point loading flexure and bending stress results were compared to cement mortar (OPC) as a reference (cement-to-sand was 1:3, W:C was 0.48 and 34.1 MPa CS at 28 days selected from trial mixtures) throughout this experiment.

Figure 6.20 demonstrates the influence of different GBFS levels on the bending stress of GPM-notched concrete substrate. The studied GPMs presented superior performance having bending stress between 4.8 and 5.9 MPa compared to cement

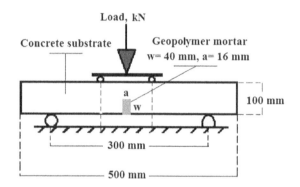

FIGURE 6.18 Schematic diagram showing the beam compatibility.

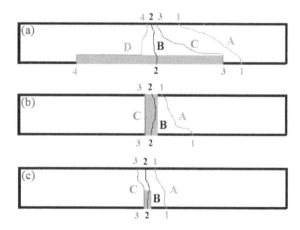

FIGURE 6.19 Compatibility evaluation according to failure mode: A, B-compatibility; C, D-incompatibility; (a) composite beam, (b) bonding beam, and (c) bending stress.

Compatibility of Geopolymer

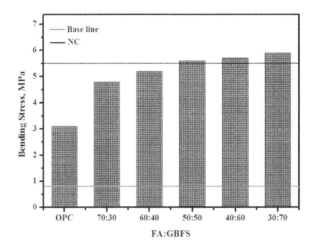

FIGURE 6.20 FA:GBFS ratio–dependent bending stress of GPM-notched concrete substrate beam.

mortar as control sample (3.1 MPa) and the baseline sample (0.7 MPa). The bending stress of the GPMs enhanced from 4.8 to 5.9 MPa with increasing GBFS level from 30% to 70%, correspondingly. The impact of GBFS ratios on failure pattern of GPMs was also examined. From visual investigation, the bending stress of cement mortar and 30% GBFS content revealed a failure type C. For 40% and 50% GBFS, the GPM displayed failure type B as opposed to failure type A for 60% and 70% of GBFS. The proposed GPMs included with high volume of GBFS (40% and above) revealed sufficiently elevated bending stresses and enhanced compatibility with the concrete substrate, which was even pronounced for the mixes containing 60% and 70% of GBFS. This observation was ascribed to the enhancement in the reaction products which in turn resulted in the improvement of strength and bonding capacity of GPM-notched NC beam containing high-volume GBFS.

6.9 SUMMARY

Depending on the evaluation of the fresh, mechanical attributes and the similarity between GPMs and concrete substrate (as repair martial), the following conclusions were drawn:

  i. High-performance GPM with low alkaline solution concentration can be obtained from waste materials. Use of GBFS (as waste material) could enhance the Ca^{++} ion concentration in the alkali-activated matrix and substitute the low Na^+ ion concentration in the geopolymerization development.
  ii. Intensity of geopolymerization can be enhanced by the inclusion of GBFS, where an increase in the calcium concentration was found to be responsible for the enhanced dissolution and precipitation of Al_2O_3 and SiO_2.
  iii. Replacing FA by 50% GBFS in the blended mix resulted in the optimum flowability of GPMs, where an increase or decrease in the GBFS content

could diminish the workability of mortars. Increasing GBFS content as a replacement of FA could reduce the initial and final setting times of GPMs.

iv. An increase in GBFS content resulted in the formation of more C-S-H and C-A-S-H gels beside N-A-S-H gel, which in turn increased the bond strength and improved the microstructures of GPMs especially after 28 days of age.

v. The achievement of excellent SSBS of the studied GPMs suggested their potential to be used as alternative repair materials for damaged concrete structures.

vi. Results on the compatibility between GPMs and concrete substrate obtained by thermal expansion coefficient, three-point loading composite beam and bending stress measurements for the mortar prepared with GBFS contents of 40% and above revealed great suitability for repairing works.

REFERENCES

1. Jiang, L. and D. Niu, Study of deterioration of concrete exposed to different types of sulfate solutions under drying-wetting cycles. *Construction and Building Materials*, 2016. **117**: pp. 88–98.
2. Mirza, J., et al., Preferred test methods to select suitable surface repair materials in severe climates. *Construction and Building Materials*, 2014. **50**: pp. 692–698.
3. Kumar, G.R. and U. Sharma, Abrasion resistance of concrete containing marginal aggregates. *Construction and Building Materials*, 2014. **66**: pp. 712–722.
4. Wang, B., S. Xu, and F. Liu, Evaluation of tensile bonding strength between UHTCC repair materials and concrete substrate. *Construction and Building Materials*, 2016. **112**: pp. 595–606.
5. Huseien, G.F., et al., Synthesis and characterization of self-healing mortar with modified strength. *Jurnal Teknologi*, 2015. **76**(1): pp. 195–200.
6. Alanazi, H., et al., Bond strength of PCC pavement repairs using metakaolin-based geopolymer mortar. *Cement and Concrete Composites*, 2016. **65**: pp. 75–82.
7. Huseien, G.F., K.W. Shah, and A.R.M. Sam, Sustainability of nanomaterials based self-healing concrete: An all-inclusive insight. *Journal of Building Engineering*, 2019.
8. Ouellet-Plamondon, C. and G. Habert, Life cycle assessment (LCA) of alkali-activated cements and concretes, in *Handbook of alkali-activated cements, mortars and concretes*. 2015, Elsevier. pp. 663–686.
9. McLellan, B.C., et al., Costs and carbon emissions for geopolymer pastes in comparison to ordinary portland cement. *Journal of Cleaner Production*, 2011. **19**(9–10): pp. 1080–1090.
10. Liu, Z., et al., Characteristics of alkali-activated lithium slag at early reaction age. *Journal of Materials in Civil Engineering*, 2019. **31**(12): p. 04019312.
11. Huseien, G.F., et al., Alkali-activated mortars blended with glass bottle waste nano powder: Environmental benefit and sustainability. *Journal of Cleaner Production*, 2019. **243**: p. 118636.
12. Wu, Y., et al., Geopolymer, green alkali activated cementitious material: synthesis, applications and challenges. *Construction and Building Materials*, 2019. **224**: pp. 930–949.
13. Shekhawat, P., G. Sharma, and R.M. Singh, Strength behavior of alkaline activated eggshell powder and flyash geopolymer cured at ambient temperature. *Construction and Building Materials*, 2019. **223**: pp. 1112–1122.
14. Li, N., et al., A mixture proportioning method for the development of performance-based alkali-activated slag-based concrete. *Cement and Concrete Composites*, 2018. **93**: pp. 163–174.

15. Huseien, G.F., et al., Geopolymer mortars as sustainable repair material: a comprehensive review. *Renewable and Sustainable Energy Reviews*, 2017. **80**: pp. 54–74.
16. Kubba, Z., et al., Impact of curing temperatures and alkaline activators on compressive strength and porosity of ternary blended geopolymer mortars. *Case Studies in Construction Materials*, 2018. **9**: p. e00205.
17. Huseien, G.F., et al., Evaluation of alkali-activated mortars containing high volume waste ceramic powder and fly ash replacing GBFS. *Construction and Building Materials*, 2019. **210**: pp. 78–92.
18. Phoo-ngernkham, T., et al., Adhesion characterisation of Portland cement concrete and alkali-activated binders. *Advances in Cement Research*, 2018. **31**(2): pp. 69–79.
19. Provis, J.L., A. Palomo, and C. Shi, Advances in understanding alkali-activated materials. *Cement and Concrete Research*, 2015. **78**: pp. 110–125.
20. Li, N., N. Farzadnia, and C. Shi, Microstructural changes in alkali-activated slag mortars induced by accelerated carbonation. *Cement and Concrete Research*, 2017. **100**: pp. 214–226.
21. Huseien, G.F., et al., Effects of POFA replaced with FA on durability properties of GBFS included alkali activated mortars. *Construction and Building Materials*, 2018. **175**: pp. 174–186.
22. Hanjitsuwan, S., et al., Strength development and durability of alkali-activated fly ash mortar with calcium carbide residue as additive. *Construction and Building Materials*, 2018. **162**: pp. 714–723.
23. Li, N., et al., Composition design and performance of alkali-activated cements. *Materials and Structures*, 2017. **50**(3): p. 178.
24. Huseiena, G.F., et al., Potential use coconut milk as alternative to alkali solution for geopolymer production. *Jurnal Teknologi*, 2016. **78**(11): pp. 133–139.
25. Huseien, G.F., et al., Effect of metakaolin replaced granulated blast furnace slag on fresh and early strength properties of geopolymer mortar. *Ain Shams Engineering Journal*, 2016. **9**(4): 1557–566.
26. Phoo-ngernkham, T., et al., Effects of sodium hydroxide and sodium silicate solutions on compressive and shear bond strengths of FA–GBFS geopolymer. *Construction and Building Materials*, 2015. **91**: pp. 1–8.
27. Huseien, G.F., et al., Waste ceramic powder incorporated alkali activated mortars exposed to elevated Temperatures: Performance evaluation. *Construction and Building Materials*, 2018. **187**: pp. 307–317.
28. Phoo-ngernkham, T., et al., High calcium fly ash geopolymer mortar containing Portland cement for use as repair material. *Construction and Building Materials*, 2015. **98**: pp. 482–488.
29. Marinković, S., et al., Environmental assessment of green concretes for structural use. *Journal of Cleaner Production*, 2017. **154**: 633–649.
30. Provis, J.L., Geopolymers and other alkali activated materials: why, how, and what? *Materials and structures*, 2014. **47**(1): pp. 11–25.
31. Chindaprasirt, P., et al., Effect of calcium-rich compounds on setting time and strength development of alkali-activated fly ash cured at ambient temperature. *Case Studies in Construction Materials*, 2018. **9**: p. e00198.
32. Yip, C.K., G. Lukey, and J. Van Deventer, The coexistence of geopolymeric gel and calcium silicate hydrate at the early stage of alkaline activation. *Cement and Concrete Research*, 2005. **35**(9): pp. 1688–1697.
33. Huseien, G.F., et al., The effect of sodium hydroxide molarity and other parameters on water absorption of geopolymer mortars. *Indian Journal of Science and Technology*, 2016. **9**(48):pp. 1–7.
34. Garcia-Lodeiro, I., et al., Compatibility studies between NASH and CASH gels. Study in the ternary diagram Na_2O–CaO–Al_2O_3–SiO_2–H_2O. *Cement and Concrete Research*, 2011. **41**(9): pp. 923–931.

35. Jang, J., N. Lee, and H. Lee, Fresh and hardened properties of alkali-activated fly ash/slag pastes with superplasticizers. *Construction and Building Materials*, 2014. **50**: pp. 169–176.
36. Palomo, A., et al., A review on alkaline activation: new analytical perspectives. *Materiales de Construcción*, 2014. **64**(315): p. 022.
37. Kürklü, G., The effect of high temperature on the design of blast furnace slag and coarse fly ash-based geopolymer mortar. *Composites Part B: Engineering*, 2016. **92**: pp. 9–18.
38. Kramar, S., A. Šajna, and V. Ducman, Assessment of alkali activated mortars based on different precursors with regard to their suitability for concrete repair. *Construction and Building Materials*, 2016. **124**: pp. 937–944.
39. Huseiena, G.F., et al., *Potential use coconut milk as alternative to alkali solution for geopolymer production*. 2016. **78**(11): pp. 133–139.
40. Davidovits, J., Geopolymer cement. A review. Geopolymer Institute, *Technical Papers*, 2013. **21**: pp. 1–11.
41. Liu, Y., et al., Compatibility of repair materials with substrate low-modulus cement and asphalt mortar (CA mortar). *Construction and Building Materials*, 2016. **126**: pp. 304–312.
42. Pattnaik, R.R. and P.R. Rangaraju, Analysis of compatibility between repair material and substrate concrete using simple beam with third point loading. *Journal of Materials in Civil Engineering*, 2007. **19**(12): pp. 1060–1069.
43. Czarnecki, L., et al., *Polymer Composites for Repairing of Portland Cement Concrete: Compatibility Project*| National Institute of Standards and Technology (NIST), Gaithersburg, Washington, 1999: pp. 1–4.
44. Memon, S.A., et al., Development of form-stable composite phase change material by incorporation of dodecyl alcohol into ground granulated blast furnace slag. *Energy and Buildings*, 2013. **62**: pp. 360–367.
45. ASTM, C., *117. Standard Test Method for Materials Finer than 75-μm (No. 200) Sieve in Mineral Aggregates by Washing*. 2003: Anl/llal Book ofIISTM S (Iards). **4**.
46. ASTM, C., *33, Standard specification for concrete aggregates*. 1994: Annual Book of Standards. **4**.
47. Salih, M.A., et al., Development of high strength alkali activated binder using palm oil fuel ash and GGBS at ambient temperature. *Construction and Building Materials*, 2015. **93**: pp. 289–300.
48. Yusuf, M.O., et al., Evolution of alkaline activated ground blast furnace slag–ultrafine palm oil fuel ash based concrete. *Materials & Design*, 2014. **55**: pp. 387–393.
49. ASTM, C., *109 standard test method for compressive strength of hydraulic cement mortars (using 2-in. or [50-mm] Cube Specimens)*. 1999: American Society for Testing and Materials. **318**.
50. Huseien, G.F., et al., Properties of ceramic tile waste based alkali-activated mortars incorporating GBFS and fly ash. *Construction and Building Materials*, 2019. **214**: pp. 355–368.
51. ASTM, A. *Standard specification for flow table for use in tests of hydraulic cement*. 2014. ASTM.
52. Testing, A.S. *Materials, standard test method for time of setting of hydraulic cement by Vicat Needle*. 2008: ASTM International.
53. Duan, P., et al., Fresh properties, mechanical strength and microstructure of fly ash geopolymer paste reinforced with sawdust. *Construction and Building Materials*, 2016. **111**: pp. 600–610.
54. Nath, S. and S. Kumar, Influence of iron making slags on strength and microstructure of fly ash geopolymer. *Construction and Building Materials*, 2013. **38**: pp. 924–930.

55. Kumar, S., R. Kumar, and S. Mehrotra, Influence of granulated blast furnace slag on the reaction, structure and properties of fly ash based geopolymer. *Journal of Materials Science*, 2010. **45**(3): pp. 607–615.
56. Sugama, T., L. Brothers, and T. Van de Putte, Acid-resistant cements for geothermal wells: sodium silicate activated slag/fly ash blends. *Advances in Cement Research*, 2005. **17**(2): pp. 65–75.
57. Puligilla, S. and P. Mondal, Role of slag in microstructural development and hardening of fly ash-slag geopolymer. *Cement and Concrete Research*, 2013. **43**: pp. 70–80.
58. Huseien, G.F., et al., Influence of different curing temperatures and alkali activators on properties of GBFS geopolymer mortars containing fly ash and palm-oil fuel ash. *Construction and Building Materials*, 2016. **125**: pp. 1229–1240.
59. Buchwald, A., H. Hilbig, and C. Kaps, Alkali-activated metakaolin-slag blends—performance and structure in dependence of their composition. *Journal of Materials Science*, 2007. **42**(9): pp. 3024–3032.
60. Pacheco-Torgal, F., J. Castro-Gomes, and S. Jalali, Investigations on mix design of tungsten mine waste geopolymeric binder. *Construction and Building Materials*, 2008. **22**(9): pp. 1939–1949.
61. Al-Majidi, M.H., et al., Development of geopolymer mortar under ambient temperature for in situ applications. *Construction and Building Materials*, 2016. **120**: pp. 198–211.
62. Shen, W., et al., Magnesia modification of alkali-activated slag fly ash cement. *Journal of Wuhan University of Technology-Materials Science Edition*, 2011. **26**(1): pp. 121–125.
63. Yu, R., P. Spiesz, and H. Brouwers, Development of an eco-friendly Ultra-High Performance Concrete (UHPC) with efficient cement and mineral admixtures uses. *Cement and Concrete Composites*, 2015. **55**: pp. 383–394.
64. Khater, H., Effect of calcium on geopolymerization of aluminosilicate wastes. *Journal of Materials in Civil Engineering*, 2011. **24**(1): pp. 92–101.
65. Myers, R.J., et al., Generalized structural description of calcium–sodium aluminosilicate hydrate gels: the cross-linked substituted tobermorite model. *Langmuir*, 2013. **29**(17): pp. 5294–5306.
66. Li, Z. and S. Liu, Influence of slag as additive on compressive strength of fly ash-based geopolymer. *Journal of Materials in Civil Engineering*, 2007. **19**(6): pp. 470–474.
67. Ranjbar, N., et al., Compressive strength and microstructural analysis of fly ash/palm oil fuel ash based geopolymer mortar under elevated temperatures. *Construction and Building Materials*, 2014. **65**: pp. 114–121.
68. Fernández-Jiménez, A. and A. Palomo, Mid-infrared spectroscopic studies of alkali-activated fly ash structure. *Microporous and Mesoporous Materials*, 2005. **86**(1): pp. 207–214.
69. Islam, A., et al., Engineering properties and carbon footprint of ground granulated blast-furnace slag-palm oil fuel ash-based structural geopolymer concrete. *Construction and Building Materials*, 2015. **101**: pp. 503–521.
70. Phoo-ngernkham, T., et al., Flexural strength of notched concrete beam filled with alkali-activated binders under different types of alkali solutions. *Construction and Building Materials*, 2016. **127**: pp. 673–678.

7 Effects of Aggressive Environments on Geopolymer Performance as Repair Materials

7.1 INTRODUCTION

Nowadays, the environmental benefits of the GPMs such as low carbon dioxide (CO_2) emissions, energy savings, reduction of landfill problems, saving of natural resources and lowering of the total demand and cost-effectiveness making them the suitable alternative materials for the traditional concretes in the civil construction sectors [1–3]. Generally, these GPMs are produced by mixing various waste materials containing high amount of the aluminosilicate (AS) with other calcium-based components using alkaline solution activation [4–6]. Diverse cheap and abundant industrial and agricultural by-products as the wastes materials such as the FA Class F and C, palm oil fuel ash (POFA), waste ceramic powder (WCP) and ground blast furnace slag (GBFS) are introduced as the main source of ASs to produce GPMs. Thus, it is realized that various industrial and agricultural waste by-products as the alkali-activated binder can efficiently be used in the GPMs to reduce the CO_2 emission level by up to 75% compared to the Ordinary Portland Cement (OPC) [7–9].

Several studies [10–13] stated that the ceramics are extremely tolerant materials against various forces or factors of the degradation. Being enriched with the crystalline ASs, these ceramics are the appropriate supplements for the cement materials that can improve the mechanical strength and durability performance of the concretes [14–16]. Despite the recycling and reusing of various kinds of the ceramic wastes, their total amount used in the construction sectors in general and concrete industries in particular is yet insignificant [13,17,18]. Therefore, the instantaneous recycling of these ceramics for other industrial applications is needed. Presently, the construction sectors need substantial amount of ceramic wastes to surmount various environmental problems. The safe use of the ceramic wastes is recommended without any significant changes in the manufacturing and applications in the free cement such as the geopolymer (GP) and alkali-activated systems. The use of ceramic wastes has many environmental implications, where the replacement of the natural resources and raw materials by these wastes can remarkably reduce the total demand and consumption, save the energy and expenditure in terms of the waste dumping in the landfill

DOI: 10.1201/9781003173618-7

thereby protecting the environment. Intensive studies showed that the building and concrete industries may get much benefits in terms of the durability and sustainability wherein the industrial and agricultural waste by-products can be reused efficiently as the practical concretes without or with OPC [19–22].

The high contents of the amorphous and crystalline form of ASs, the plentiful and low cost on the earth make FA the most widespread resource material for the fabrication of the GPs [23]. In the developed countries, some million tonnes of coals are consumed annually for the generation of electric power [24]. Therefore, the recycling of these wastes for making GPs and alkali-activated systems may directly solve the environmental problems, thereby bringing sustainability in the concrete industries [25,26]. It is important to note that the GBFS as waste is usually acquired from the molten iron slag quenching (a spin-off of iron and steel fabrication) inside a blast furnace in the presence of water. In this procedure, some kinds of granular amorphous products are obtained which is further dried and crushed into a fine powder to get GBFS [27]. Depending on the initial resources utilized for the iron manufacturing, the chemical compositions of the slag may vary significantly [28]. Because of the high CaO and SiO_2 contents, the GBFS shows both cementing and pozzolanic characteristics [29]. Earlier, the GBFS was intensively utilized for the construction purposes to enhance the durability and compressive strength (CS) of the traditional concretes [30]. The alkali-activated slag was found to exhibit very high durability [31,32], workability and strength performance.

The surfaces of the concrete structures, such as the sidewalks, pavements, parking decks, bridges, runways, canals, dykes, dams and spillways, deteriorate progressively due to a variety of physical, chemical, thermal and biological processes. Actually, the performance of concrete compositions is greatly affected by the improper usage of substances, and physical and chemical conditions of the environment [33]. The immediate consequence is the anticipated need of maintenance and execution of repairs [34]. Approximately few million tonnes of solid wastes in the form of spin-offs from agricultural and other industries (POFA, bottom ash, ceramic tiles, FA and GBFS) are discarded annually as landfill in Malaysia [35,36]. Such waste results in serious ecological problems in terms of air pollution and leaching out of the harmful products. Several researches showed the feasibility of recycling these wastes to get novel concretes as a substitute to the OPC (above 60%) [37,38]. Furthermore, such new types of concretes due to their green chemistry are environmentally friendly, durable and cheap characteristics suitable for construction materials. So far, the progress of diverse GPMs (containing the above-mentioned wastes) as repair materials especially for deteriorated concrete surfaces is rarely explored.

Morgan et al. [39] acknowledged that the compatibility between the concrete substrate and repair mortars must meet certain requirements such as the CS, tensile strength, flexural strength and bond strength. However, the bonding strength among the concrete substrate and the repairing system [40–42] decides the binding efficiency of the GPMs as repair materials. Durability of the repair materials is characterized by their tolerance against declination. A durable mortar with almost free of porosity reveals strong resistance against the sulphate and chloride attack, abrasion and high tolerance against the aggressive environmental conditions [43].

Aggressive Environmental Effects on GP

In this chapter, the effect of aggressive environment on the CS, flexural strength and the slant shear bond strength between the normal cement (NC) and prepared GPMs was assessed. These GPMs were prepared by incorporating three waste materials including the POFA, GBFS and FA or WCP, GBFS and FA (called the ternary blend). These blends were designed at various levels of POFA or WCP, GBFS and FA to determine the viability to reuse the solid wastes from diverse industries. The idea is to turn these wastes into environmentally responsive and long-lasting repairable mortars/binders for sustainable development. In this procedure, the WCP was kept in high volume (between 50% and 70% by weight) and substituted by different contents of GBFS and FA in the practical operational range with appropriate physical conditions needed to fabricate such WCP–GBFS–FA ternary mortars with the alkali activation. All the proposed mixes were analysed using different analytical techniques to assess slant shear bond stability performances (between NC and GPMs) under aggressive environmental conditions including the sulphuric acid and high heating. The experimental findings were analysed, interpreted and validated to determine the optimal compositions.

7.2 GEOPOLYMER TERNARY BLENDED

In this work, the starting materials such as the POFA, WCP, GBFS, FA, river sand, sodium hydroxide (NH) and sodium silicate (NS) were utilized. The materials including the POFA or WCP, FA and GBFS were stored in an air-tight plastic storage box to avoid any contamination. The ternary components of POFA or WCP, GBFS and FA were mixed to develop the GPMs as binders. The materials such as the GBFS and FA were utilized as received but the POFA and WCP were treated in the laboratory prior to the use. The GP binder was prepared using the pure GBFS as one of the source materials, which was used without any treatment in the laboratory. The Class F FA (low content of calcium) was collected from the power station as the source of ASs to make GPMs. POFA is a waste material produced from the palm oil fibres, bunches and shells as fuel for power generation in the mills. It is obtained from Kilang Sawit PPNJ Kahang located in Johor (Malaysia). This POFA is collected from an ash outlet kept away from the boiler burning chamber. It is first sieved with 600 µm to remove out the large particles then dried in the oven at 105°C for 24 hours to remove free water. This is followed by sieving with 300 µm to screen large particles for improving the efficiency of the ash. The sieved ash passing from 300 µm sieve is then ground to fine particles for 6 hours using Los Angeles Abrasion machine that operates with 15 stainless balls each of diameter 50 mm and drum speed in the range of 32–35 revolutions per minutes (rpm). The duration of grinding influences the fineness of particles which is monitored at every 1-hour interval. The percentage of particles retained on 45 µm sieves is reduced with grinding duration. After complete grinding, all the ash passes through the sieve (45 µm). The WCP was obtained from the White Horse ceramic manufacturer as waste materials. The wastes from the homogeneous ceramic tiles only (with identical thickness and without glass coating) were acquired to make the binders. First, the collected WCP was ground in a crushing machine before being sieved through 600 µm to isolate the large particles. Again, the sieved WCP was ground for 6 hours using Los Angeles Abrasion machine to get

TABLE 7.1
The Chemical Composition and Physical Properties of Raw Materials

Materials	POFA	WCP	GBFS	FA
Chemical compositions (%)				
SiO_2	64.20	72.64	30.82	57.20
Al_2O_3	4.25	12.23	10.91	28.81
CaO	10.20	0.02	51.82	5.16
Fe_2O_3	3.13	0.56	0.64	3.67
Na_2O	0.10	13.46	0.45	0.08
MgO	5.90	0.99	4.57	1.48
K_2O	8.64	0.03	0.36	0.94
Loss on ignition	1.73	0.13	0.22	0.12
Physical properties				
Specific gravity	1.96	2.61	2.9	2.20
Surface area-BET (m^2/g)	23.1	12.2	13.6	18.1
Mean diameter (μm)	8.2	35	12.8	10

the desired particle size of 66% passing through 45 μm in accordance to the ASTM 618 [44]. Next, the powder was obtained and utilized to make the mortar mixes. Table 7.1 summarizes the chemical composition and physical characteristics of the resource materials in terms of their particle sizes, specific gravities and surface areas obtained from the laser diffraction particle size analyser (PSA, Mastersizer, Malvern Instruments) and Brunauer–Emmett–Teller (BET) test. The POFA, WCP, GBFS and FA particles size distributions were achieved by PSA with the median values of 8.2, 35, 12.8 and 10 μm, respectively.

The X-ray fluorescence (XRF) spectra of the prepared mixes and raw materials (POFA, WCP, GBFS and FA) were recorded to verify their chemical compositions where a wavelength-dispersive XRF spectrometer was used (Quant Express dry method interfaced with Spectra Plus software). For the XRF analysis, the samples were prepared using the fused-bead method. Furthermore, the time of hardening was monitored using a Vicat apparatus and the moulds were checked for every 5 min intervals to assess the status of hardening. The chemical composition of raw materials (POFA, WCP, GBFS and FA) determined from the XRF test is summarized in Table 7.1. The silica and aluminium are found to be the main oxides in the POFA (68.5%), WCP (84.8%) and FA (86%) composition compared to 41.7% in the GBFS. The WCP revealed the high concentration of silicates (72.6%) and GBFS showed a very high level of CaO (51.8%). Regarding the aluminium oxide content, the FA presented the highest level (28.8%) than the GBFS, POFA and WCP. Besides the amount of the silicates, Al and Ca oxides played a vital role in the fabrication of the GPMs wherein the dense gels including the calcium–silicate–hydrate (C-S-H), aluminate-substituted calcium–silicate–hydrate (C-A-S-H) and sodium–aluminium–silicate (N-A-S-H) were formed by the geopolymerization process. In WCP chemical

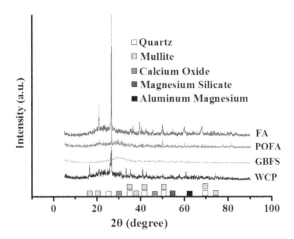

FIGURE 7.1 The XRD patterns of the WCP, GBFS, POFA and FA.

composition, the mass ratio of sodium oxide (Na_2O) was much higher (13.5%) than that of in GBFS (0.45%), POFA (0.1%) and FA (0.08%). The loss on ignition (LOI) contents were found to be very less in the WCP (0.13%), GBFS (0.22%) and FA (0.12%) compared to POFA (1.73%).

The X-ray diffraction (XRD) patterns of the POFA, WCP, GBFS and FA are shown in Figure 7.1. The XRD analyses revealed the presence of low calcium content in POFA and WCP, and high calcium content in FA with the sharp diffraction peaks around $2\theta = 16°–30°$, which were assigned to the crystalline lattice planes of the silica and alumina compounds. Nevertheless, the other characteristic peaks were attributed to the existence of the crystalline phases of quartz and mullite. Several studies [25,32,45] also showed that the presence of amorphous phases of WCP and FA plays a major role on hydration development and gel formation. In contrast to WCP and FA, GBFS showed a strong glassy character in the absence of any prominent diffraction peak. The existence of the reactive amorphous silica and calcium at high concentration in GBFS was highly prospective for the GPM synthesis. In fact, the inclusion of FA and GBFS was needed to surmount little Al_2O_3 (12.2 wt.%) and CaO (0.02 wt.%) concentration that was present in the WCP. From the scanning electron microscope (SEM) results, it was found that both WCP and GBFS comprised angular particles with rough surface, whereas the FA was composed of spherical particles with smooth surface. POFA revealed that weathered particles from ground are spherical-shaped particles that resulted in the combination of spherical and irregular-shaped particles.

In this experiment, the natural river sand was utilized as fine aggregates to fabricate all the GPMs. Following the ASTM C117 standard, the collected river sand was first washed using water to remove the silts and impurities. Next, the cleaned river sand was oven-dried for 24 hours at 60°C to minimize the moisture content during the fabrication of GPMs. The specific gravity of the treated river sand was assessed and estimated with the value of 2.6. For the preparation of mortar specimens, analytical-grade NaOH with purity of 98% was used to make the alkaline activator solution.

First, the NaOH pellets were dissolved in water to make 4 M of NH solution. Then, the resultant NH solution was cooled for 24 hours before being added with Na_2SiO_3 solution to achieve the final alkali mixture with modulus of 1.02 ($SiO_2:Na_2O$). In the ultimate prepared alkaline solution, Na_2SiO_3:NaOH ratio (NS:NH) was maintained to be 0.75 for all GPM mixes. The low content of Na_2SiO_3 and molarity of NaOH (4 M) were considered to reduce the cost, energy consumption, CO_2 emission, hazard and environmental effects of the alkaline solution, thereby increasing the sustainability of GPMs compared to other studies [46,47] that used NS:NH of 2.5 and up to 12 M of NaOH.

Various ternary blends were prepared using the raw materials including the POFA, WCP, GBFS and FA to produce GPMs and the details of the mixes are illustrated in Table 7.2. The level of WCP was kept high (between 0.191 and 0.268 m^3 by volume) for all the designed mixes. However, the level of GBFS in the GPMs mixes was maintained between 0.068 and 0.171 m^3 (by volume) with the replacement of the FA powder at various levels. For continuity, all the mixtures had equal levels of the NH molarity (4 M) with 0.384 m^3 fine aggregates for each batch. The ratio of the binders (such as the POFA or WCP, GBFS and FA) to the fine aggregates (B:A) were kept in the range of 0.94%–1.03%. For all the GPMs mixes, the fixed contents of the alkali activator solution including the sodium silicate (Na_2SiO_3) and sodium hydroxide (NaOH) were 171.43 and 228.57 kg, respectively. The prepared GPMs were characterized by different tests to determine the impact of WCP at elevated volume on their bond strength performance and durability.

TABLE 7.2
The Compositions and Proportions of Various Components Used for the GPMs Synthesis

Mix	Binder, Weight %				Fine Aggregates	Alkaline Solution	
	GBFS	FA	WCP	POFA		NaOH	Na_2SiO_3
$GPMs_1$	50	0	50	0	1.0	0.228	0.171
$GPMs_2$	40	10	50	0	1.0	0.228	0.171
$GPMs_3$	30	20	50	0	1.0	0.228	0.171
$GPMs_4$	20	30	50	0	1.0	0.228	0.171
$GPMs_5$	40	0	60	0	1.0	0.228	0.171
$GPMs_6$	30	10	60	0	1.0	0.228	0.171
$GPMs_7$	20	20	60	0	1.0	0.228	0.171
$GPMs_8$	30	0	70	0	1.0	0.228	0.171
$GPMs9$	20	10	70	0	1.0	0.228	0.171
$GPMs_{10}$	50	50	0	0	1.0	0.228	0.171
$GPMs_{11}$	50	40	0	10	1.0	0.228	0.171
$GPMs_{12}$	50	30	0	20	1.0	0.228	0.171
$GPMs_{13}$	50	20	0	30	1.0	0.228	0.171
$GPMs_{14}$	50	10	0	40	1.0	0.228	0.171
$GPMs_{15}$	50	0	0	50	1.0	0.228	0.171

7.3 PROCEDURES OF GEOPOLYMER TESTS

The cubic specimens (dimension of 50 mm × 50 mm × 50 mm) of the prepared mortars were subjected to the hardening tests (CS) following the ASTM C579 standard. In accordance to the ASTM C 109, the CS test was made with the load rate of 2.5 kN/s and a minimum of three specimens were used to assess the average CS at different curing periods (1, 3, 7, 28, 56, 90, 180 and 360 days). The capacity of the slant shear bonding among the NC substrate and different GPMs with the stiffer slant shear angle of 30° were determined in accordance with the ASTM C882 [48]. For the sample preparation, the NC was first cast in the cylinder moulds (dimension of 100 mm × 200 mm) and cured in water for 3 days. Next, these samples were left at 25°C ± 3°C with the relative humidity of 75% until 28 days of age. Afterwards, these specimens were cut into half following the half-slanted dimension (with 30°) before being fixed in the cylindrical moulds (dimension of 100 mm × 200 mm) (Figure 7.2). It was cast for the second part (OPC and GPMs) to evaluate their slant shear bond strength (defined as the ratio of the maximum load at the failure to the bond area) performance at different ages of the curing (1, 3, 7 and 28 days). The average of the three samples from every GPM mix was considered to report the results of the slant shear bond strength.

The evaluation of the bond strength performance of the GPMs under the elevated temperatures is one of the main objectives of this study. The specimens were tested using the automatic electric furnace. The cubical specimens of the GPMs (dimension of 50 mm × 50 mm × 50 mm) were cast according to the ASTM C597 and cured for 28 days in the ambient temperatures. Form each set of the GPMs, a total of three samples were tested at high temperatures (400°C, 700°C and 900°C) under various time durations (Figure 7.3). The GPMs were cooled by the air cooling method after heating. The slant shear bond strength before and after the high temperature exposure was measured to determine the elevated temperature–mediated bond loss.

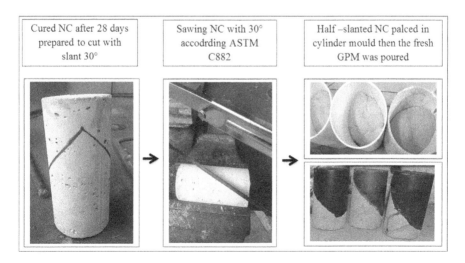

FIGURE 7.2 Procedure for the slant shear bond strength test of the GPMs.

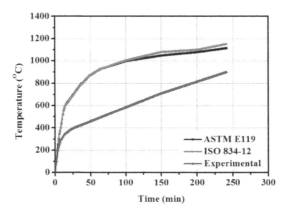

FIGURE 7.3 Time against temperature variation for ASTM E119 [49] and ISO 834-12 [50].

Finally, the relative quality of the GPMs was assessed after heating. In addition, the XRD and SEM analyses of the GPM specimens were performed to determine their microstructures and surface morphologies.

The concretes and mortars exposed to the sulphuric acid (H_2SO_4) are known to lose their strength rapidly and undergo structural deterioration. To perform the acid attack test of the binder matrices, the H_2SO_4 solution (with 10% concentration) was prepared using deionised water. The impact of the H_2SO_4 exposures on the GPMs was evaluated where six specimens for each mixture were selected at 28 days of curing. The slant shear bond strength was also assessed before immersing in the acidic medium. Each GPM mix was engrossed in the acid solution for 1 year, and the solution was altered every 90 days to maintain its pH all throughout the test period. In each instance (half and 1 year), the acid-exposed GPMs were monitored to evaluate their performance depending on the qualitative examination and residual bond strength according to the ASTM C267 (2012) standard. These acid attacks on the GPMs were basically due to the transport of the sulphate ions $(SO_4)^{2-}$ together with calcium, magnesium or sodium cations into the mortars at different concentrations. Sulphate attack on the studied mortar specimens was essentially caused by the sulphate ions $(SO_4)^{2-}$ that were transported into the mortar from varying concentrations in water along with calcium, magnesium or sodium cations. Magnesium sulphate ($MgSO_4$) solution was used to assess the resistance of the GPM specimens to sulphate attack following the same procedure as used for sulphuric acid attack test.

7.4 COMPRESSIVE STRENGTH PERFORMANCE

Figure 7.4 shows the influence of POFA-to-FA ratio on the CS of the ternary GPMs containing 50% of GBFS. The gain in the CS was monotonically increased with the increase in age. For all mixtures, the CS development rate continued to increase after 28 days and showed a percentage increase higher than 108%. After 360 days, the CS showed more than 3% increase with increasing hydration time as compared to 28 days. An increase in the POFA content was found to reduce the early age strength, as the POFA content increased from 0% to 50%. At the early age after 24 hours, the

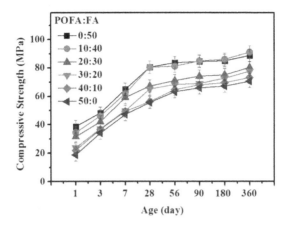

FIGURE 7.4 Effect of POFA-to-FA ratio on the CS of GPMs containing GBFS.

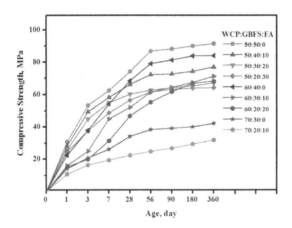

FIGURE 7.5 CS of GPMs tested in different periods.

CS revealed a decrease from 38.7 to 18.8 MPa; a percentage loss of more than 50%. For the GPM containing FA as a replacement by 10% of POFA and 50% of GBFS at ages of 90, 180 and 360 days enhanced the CS by 0.5, 1.2 and 2.6, respectively. Ranjbar et al. [51] reported that an increase in silicate-to-aluminium ratio produced a negative effect on the CS of GPM. Moreover, a reduction in Al content with increasing POFA content showed a significant influence on the C-A-S-H product with lower strength and slower chemical reaction rates. According to Ariffin et al. [52], low content of Al_2O_3 could be responsible for the reduced CS. This observed reduction in the CS was attributed to the incomplete geopolymerization process where Al_2O_3 revealed higher rate of dissolution during the early stage of geopolymerization.

Figure 7.5 displays the influence of the high WCP contents on the early and late CS development of the studied GPMs. The increment in the CS was found to be directly proportional to the curing time of all the GPM specimens. The CS of the GPMs was found to vary inversely with the increase in the WCP content from 50% to 70%, wherein the corresponding CS was decreased from 70.1 to 34.8 MPa at 28 days. The reduction in total calcium content and increment in the silica content with an increase in the content of WCP in the GPMs were responsible for this drop [22,53,54]. This negatively affected the production of C-S-H and C-A-S-H gels and thus lowered the CS of GPMs. This reduction was majorly ascribed to the high content of the silica (70%) and larger particle size (35 μm) of the WCP. Moreover, the GPM cast with high WCP contents produced the CS values of 81%, 94% and 97% at 28, 56 and 90 days of age, respectively. Conversely, the content of silica and aluminium was increased in GPM matrix with an increase in the FA replacement for GBFS which negatively affected the calcium level. Thereafter, the level of GBFS in $GPMs_1$ was decreased from 50% to 20% and the loss of CS in $GPMs_{10}$ was more than 70%, indicating a drop in the respective CS from 84.6 to 24.8 MPa at 28 days of age.

Compared to 28 days, the CS at the late age (365 days) was generally found to enhance with the increase in the hydration time for all the GPMs. The development of low-quantity C-(A)-S-H gels was attributed to the little contents of Ca which was responsible for the strength loss of studied GPMs. Rashad [55] also reported similar diminish in the CS value of FA-incorporated mortars. Actually, the observed loss in the CS may arise due to many reasons. First, the dissimilarities of chemical composition among the WCP, GBFS and FA which considerably influenced the alkaline solution activation into the binder matrix. Second, compared to GBFS, WCP and FA have poorer reaction rate due to their partial dissolution [56]. Third, with the increase in the WCP and FA levels, the compactness and density of the GPM matrix might have reduced. Fourth, the low contents of the NH (4 M) where the CS was majorly decided by the CaO content for replacing the low amount of Na_2O. In short, the generation of more C-S-H and C-A-S-H gels together with the N-A-S-H gel could enhance the CS of GPMs.

7.5 BOND STRENGTH OF GEOPOLYMER

The results of the slant (for 30°) shear bond strength tests of the GPMs containing high volume of WCP at the 1, 3, 7 and 28 days are illustrated in Figure 7.6. The bond strength is increased with the time of ageing. However, the bond strength of GPMs was dropped with the increasing content of WCP replaced by GBFS. First, the bond strength of GPMs containing 50% of WCP at 28 days was 4.2 MPa and then dropped from 3.8 to 2.7 MPa with the rise in the WCP level from 60% to 70%, respectively. Similarly, the bond strength of all GPMs was higher than that of the normal OPC mortar. It is evident that the GPMs possessed excellent bonding character than the cement materials. The effects of FA replacement for GBFS in the GPM matrix were also evaluated and the bond strength was reduced with the rise in the FA level. Furthermore, the bond strength was dropped from 4.2 to 2.7 MPa with the increase in the FA level from 0% to 30%, respectively (for specimens prepared with 50% WCP). Similar trend was observed for the mortar specimens prepared with 60%–70% of the WCP wherein the rise in the FA contents resulted in the reduction of bond strength. This is in line with the previous report [57,58] on the improved bond strength of AS-based GPs, in which, with the increase in the calcium content, the extra C–(A)–S–H gel was co-existed with the N-A-S-H gel. The increase in

Aggressive Environmental Effects on GP   133

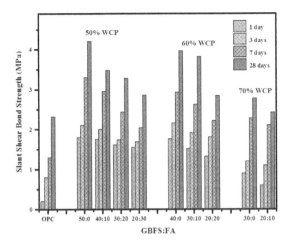

FIGURE 7.6 The 30° slant bond strength of GPMs containing various levels of WCP.

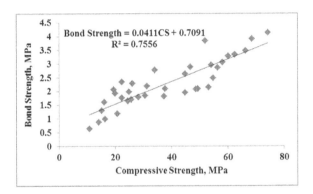

FIGURE 7.7 Relationship between the CS and bond strength.

the reaction products at the interface transition zone between the NC substrate and GPMs was responsible for the enhanced strength at the contact zone [59]. However, the GPMs with a high value of WCP content (70%) exhibited a slight decrease in the shear bond strength.

The results for the CS and slant shear bond strength of GPMs at the ages of 1, 3, 7 and 28 days are presented in Figure 7.7. The bond strength values were found to be correlated with their CS. The achieved CS values were used as a response factor with the bond strength values for predictive parameters. The linear regression analyses were performed to relate the experimental data using Eq. (7.1). For all samples, the value of R^2 was found to be 0.76, signifying excellent correlation confidence. The linear regression relation can be written as follows:

$$Bond\ Strength = 0.0411CS + 0.7091\ (R^2 = 0.7556) \tag{7.1}$$

7.6 EFFECT OF SULPHURIC ACID ATTACK

Figure 7.8 shows the influence of POFA-to-FA ratio on the residual CS of GPMs after immersing them in 10% of H_2SO_4 solution for 6 and 12 months. The resistance of the proposed GPMs against sulphuric acid attack was remarkably enhanced, wherein an inverse relationship between the residual CS and POFA content was observed. When the content of POFA was increased from 0% to 50%, the residual CS after 12 months of immersion was correspondingly reduced from 57% to 37%. The weight loss values were also decreased with the increase in POFA content (Figure 7.9).

Figure 7.9 presents the POFA content–dependent weight loss of GPMs after immersing in 10% H_2SO_4 solution for 6 and 12 months. The weight loss of alkali-activated mortar (AAM) was decreased from 1.1% to 0.1% with the increased POFA level from 0% to 50%, respectively, after 12 months of immersion in the acid solution. Mortars with higher POFA level revealed superior resistance to sulphuric acid attack compared to the one containing lower amount of POFA. With the increase in POFA level from 0% to

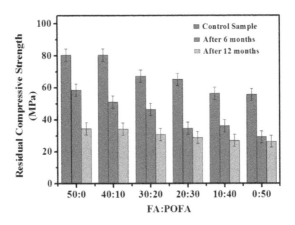

FIGURE 7.8 POFA content–dependent variation in the residual CS of GPMs after immersing in 10% of H_2SO_4 solution.

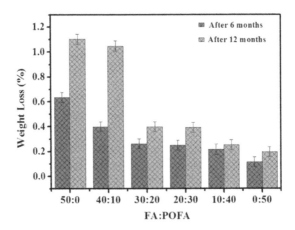

FIGURE 7.9 Weight loss of GPMs after immersing them in 10% H_2SO_4 solution.

50%, the silica content in the AAM was increased from 44% to 47.5% and aluminium content was decreased from 19.8% to 7.6%. This in turn could affect the formation of C-A-S-H gels and subsequent improvement in the specimens' strength.

Figure 7.10 shows the effect of POFA replacement by FA on the ultrasonic pulse velocity (UPV) of GPMs after immersing in 10% of H_2SO_4 solution for 6 and 12 months. Irrespective of the duration of immersion, the UPV values were decreased with the increase in POFA incorporation into the GPMs.

The XRD patterns of GPM specimens immersed in 10% of sulphuric acid (H_2SO_4) solution for 360 days are presented in Figure 7.11. For AAM with 0% POFA, the main phases detected were still evident in the GPM samples in addition to albite, gmelinite,

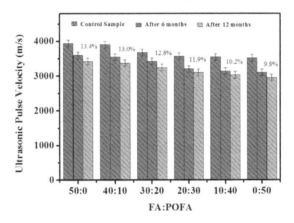

FIGURE 7.10 Impact of POFA replacement by FA on UPV of GPMs after immersing them in 10% H_2SO_4 solution for 6 and 12 months.

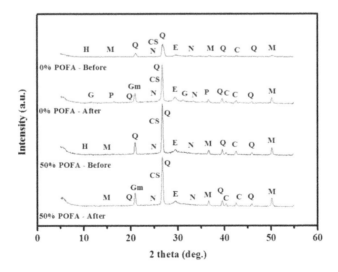

FIGURE 7.11 XRD of GPMs exposed to 10% sulphuric acid solution. M: Mullite, Q: SiO_2, CS: C-S-H gel, H: Hydrotalcite, C: Calcite, N: Nepheline, E: Anorthite, G: gypsum, Gm: gmelinite, P: portlandite.

gypsum and portlandite. It can be clearly seen that the peak intensities were significantly sharp for quartz (SiO_2), especially at 26.8°, 40° and 46.2° (2θ degrees) compared to XRD results before immersion. Gypsum ($CaSO_4 \cdot 2H_2O$) appeared in the GPMs containing 0% POFA as a new peak at $2\theta = 12.8°$ and 31.2°. At $2\theta = 21°$, the gmelinite peak was also detected. For AAM incorporating 50% POFA, less change in peak intensity was observed in specimens before and after they were immersed in acid solution. The gmelinite peak was detected as a new peak at $2\theta = 21°$. According to the XRD results, it is indicated that replacing FA by POFA shows a better resistance to sulphuric acid attack. This could be related to the rate of deterioration of sulphuric acid on GPMs.

The microstructures of the matrices near the external surface of mortars after 360 days of immersion in 10% H_2SO_4 solution were studied using SEM and the results are shown in Figure 7.12. The SEM of AAM containing 0% POFA (Figure 7.12a) showed more cracks compared to 50% POFA specimens and the presence of gypsum and gmelinite in the samples.

The results of slant shear bond strength tests at 30° for the GPMs exposed to acid environment for 180 and 365 days are illustrated in Figure 7.13. For all the samples, the bond strengths were decreased with the increased time duration of exposure in the acid solution. However, the loss in the bond strength of the GPMs was decreased with the increasing content of WCP replaced by GBFS. The loss in the bond strength of the GPMs containing 50% of WCP after 180 days of exposure was 23.7% which was dropped from

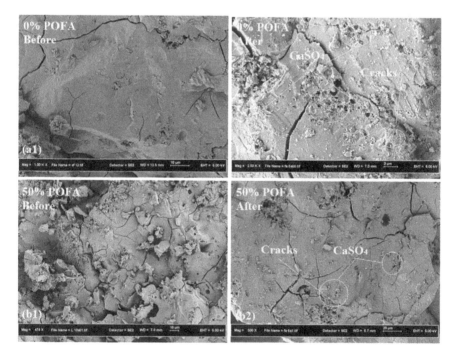

FIGURE 7.12 SEM images of 0% and 50% POFA of GPMs before and after 360 days of immersion in 10% H_2SO_4 solution.

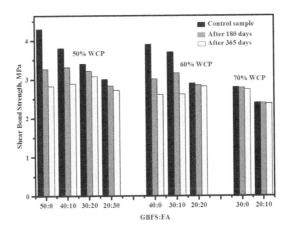

FIGURE 7.13 Effect of acid attack on shear bond strength of GPMs.

19.8% to 1.3% with the respective increase in WCP from 60% to 70%. Similarly, the loss in the bond strength of GPMs after 365 days was dropped from 34.1% to 3.1% with the increase in WCP content from 50% to 70% as a replacement of GBFS matrix, respectively. It is evident that the GPMs with high content of WCP possessed excellent bonding character in the acid environment than the GBFS materials. The effect of FA substitution in GBFS on the bond strength of GPMs at various levels of WCP is evaluated. It was demonstrated that a rise in the FA level to 30% for the mortar matrix prepared with 50% (high volume) of WCP can lower the loss in the bond strength around 18% compared to the specimen prepared without FA. However, the specimens prepared with the elevated quantity of WCP (70%) and 10% of FA presented an excellent performance and showed loss in the bond strength lower than 0.7% and 1.1% after 180 and 365 days, respectively. When mortar specimens were exposed to sulphuric acid, $Ca(OH)_2$ reacted with SO_4^{-2} and formed gypsum ($CaSO_4.2H_2O$). This in turn resulted in matrix expansion and extra cracks within the specimens [60] thereby led to the bond strength weakening between the mortar and NC. The reduction in GBFS content replaced by WCP and FA in the mortar matrix led to restrict the amount of $Ca(OH)_2$ and reduced the formation of gypsum and cracks. This disclosure attributed to the excellent performance of the GPMs prepared with high contents of WCP and FA.

Figure 7.14 displays the XRD patterns of the proposed GPMs under 10% of H_2SO_4 exposure (immersion) after 360 days. These GPMs showed the existence of gypsum (calcium sulphate hydrate). The intensity of gypsum peak at $2\theta = 29.8°$ was reduced with the rise in FA contents. In addition, a gypsum peak at 11.8° and another at 20.9° near to the quartz peak (20.8°) were appeared. Bellmann and Stark [61] also reported the appearance of quartz and gypsum peaks at 20.8° and 20.9° 2θ, respectively, which were hard to differentiate. Nevertheless, double peaks that were evidenced upon the close scrutiny indicated the attendance of both quartz and gypsum. An increase in the FA level to 30% in place of GBFS in the mortar could restrict the formation of gypsum as indicated by the appeared peaks at 29.8°, thereby enhancing the resistance of mortars against the H_2SO_4 attack. In addition, it was suggested that

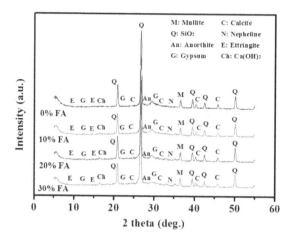

FIGURE 7.14 XRD patterns of GPMs containing 50% of WCP and exposed to H_2SO_4 environment for 12 months.

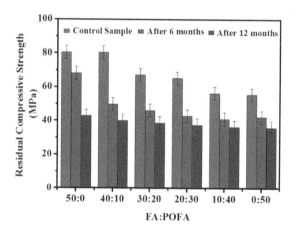

FIGURE 7.15 Residual CS of GPMs after immersing them in 10% $MgSO_4$ solution.

GPMs containing high volume of GBFS can generate more C-A-S-H and N-A-S-H gels that are susceptible to the H_2SO_4 attack.

7.7 GEOPOLYMER RESISTANCE TO SULPHATE ATTACKS

Figure 7.15 shows the residual CS of GPMs after immersing them in 10% $MgSO_4$ solution for 6 and 12 months compared to the control sample. Irrespective of the immersion period, the residual CS of the designed mixes was reduced with the increase in POFA contents. The residual CS and change in weight of the proposed GPMs were measured to evaluate the resistance against sulphate attack. Thus, all the specimens with POFA content above 10% presented higher resistance to sulphate attack compared to the specimens with high FA content. An increase in the POFA content from 10% to 50% led to a decrease in the respective residual strength from 65% to 53%.

Figure 7.16 displays the weight loss of GPMs after immersing them in 10% $MgSO_4$ solution for 6 and 12 months. Irrespective of the immersion period, the weight loss of the designed mixes was influenced marginally with the increase in POFA contents. Also, the deterioration of alkali-activated specimen's surface and weight loss was decreased with the increase in POFA content. The weight loss was decreased from 0.62% to 0.58% with the increase in POFA from 0% to 50%, respectively. Most of the researchers attributed the sulphate attack resistance of mortars to the formation of expansive ettringite ($3CaO \cdot Al_2O_3 \cdot 3CaSO_4 \cdot 32H_2O$) and gypsum [calcium sulphate dihydrate ($CaSO_4 \cdot 2H_2O$)]. This could be accompanied by the expansion or softening of alkali-activated mortars thus decreasing the CS of specimens. Furthermore, the increased POFA level in the mortar could lead to a decrease in the formation of C-A-S-H gel together with $CaSO_4$ product.

Figure 7.17 depicts the POFA content–dependent variation in the UPV of GPMs after immersing them in 10% of $MgSO_4$ solution for 6 and 12 months compared to

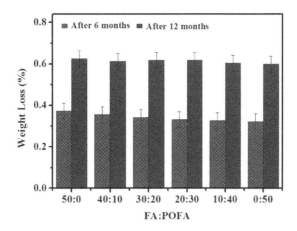

FIGURE 7.16 Weight loss of GPMs after immersing them in 10% $MgSO_4$ solution.

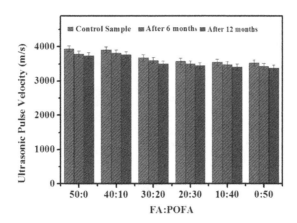

FIGURE 7.17 POFA content–dependent UPV of GPMs after immersing them in 10% $MgSO_4$ solution.

the control sample. Irrespective of the immersion periods, the UPV of the designed ternary mixes was marginally reduced with the increase in POFA contents.

Figures 7.18 and 7.19 show the results of XRD and SEM of GPMs exposed to sulphuric acid for the immersion time of 360 days. Specimens prepared with 0% POFA showed a large difference between the intensities of peaks before and after exposure to sulphate attack. Furthermore, the XRD and SEM results revealed the presence of elongated crystalline structures of gypsum ($CaSO_4 \cdot 2H_2O$). New peaks were detected such as gmelinite, albite and ettringite at 21°, 31° and 32.4° (2θ), respectively. The high

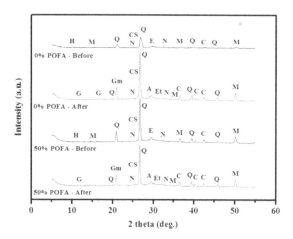

FIGURE 7.18 XRD of GPMs exposed to sulphuric acid attack. M: Mullite, Q: SiO_2, CS: C-S-H gel, H: Hydrotalcite, C: Calcite, N: Nepheline, E: Anorthite, G: gypsum, Gm: gmelinite, A: albite, Et: Ettringer.

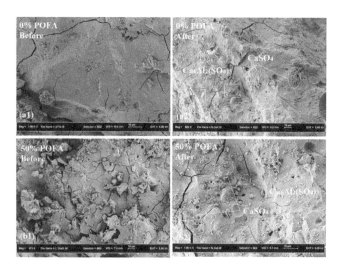

FIGURE 7.19 SEM images of GPMs containing 0% and 50% POFA before and after 360 days of immersion in 10% $MgSO_4$ solution.

rate of gypsum and ettringite formation in specimens of 0% POFA led to more cracks and showed lower resistance to sulphate attack compared to specimens of 50% POFA.

7.8 EFFECT OF ELEVATED TEMPERATURES

Figure 7.20 shows the effect of high elevated temperatures on the slant shear bond strength of GPM specimens to concrete substrate prepared with 30°. For all the GPMs, the loss in bond strength increased with the increase in temperature from 27°C to 900°C. The results indicated an increase in the loss of bond strength with the rise in temperature from 27°C to 900°C. The bonding strength at 27°C was higher for all mixtures after the exposure to various elevated temperatures. The loss in the bond strength values was decreased with the addition of WCP and FA, where the loss percentage at 400°C was dropped from 37% to 3.8% with the addition of WCP in place of GBFS from 50% to 70%, respectively. The effect of FA substitution in GBFS in each level of high-volume WCP on the slant shear bond strength was also assessed. The positive effect on the bond value was observed with the rise in the FA content. The percentage of loss on the bond strength was dropped from 37% to 21% with the respective increase in FA substitution for GBFS from 0% to 30%. Similar trend was observed in the specimens exposed to 700°C and 900°C. Those prepared with the high volume of WCP and FA showed lower loss in the bond strength compared to other samples. However, the highest loss on the bond strength (82%) was observed in GPMs made from 50% of WCP and GBFS compared to the mortar (41%) that contained 70% of WCP, 20% of GBFS and 10% of FA under the exposure of 900°C.

The XRD patterns of synthesized GPMs under the exposure of elevated temperatures up to 900°C are presented in Figure 7.21. The semi-crystalline alumina-silicates gel and quartz (Q) peak (Figure 7.21a) was observed after the exposure below 400°C. The appearance of broad peak in all the GPMs in the range of $2\theta = 24.8°–31°$ was ascribed to the formation of crystalline zeolites as secondary reaction products after the completion of fire resistance test. The XRD patterns of the GPMs exposed to 700°C and 900°C are depicted in Figures 7.21b and c, respectively. For WCP-based

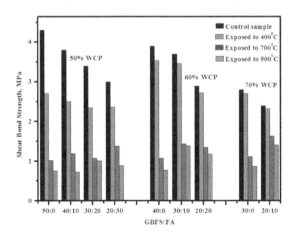

FIGURE 7.20 Effect of acid attack on shear bond strength of GPMs.

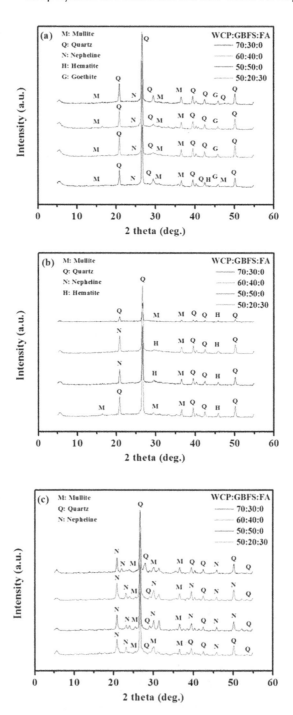

FIGURE 7.21 Specimen XRD analysis after fire exposure to different temperatures. (a) 400°C. (b) 700°C. (c) 900°C.

GPMs heated at 700°C, the emergence of the strong peaks can primarily be assigned to the presence of the mullite, quartz and nepheline phases. The mullite was the only stable crystalline structure of the Al_2O_3–SiO_2 coordination. Due to their high stability, low thermal expansion and excellent oxidation resistance against elevated temperature, mullite was retained in the mortars. After exposing to 700°C, the quartz peaks were steady but the peaks of mullite appeared increasingly intense. Furthermore, around 400°C a phase transformation from the goethite to hematite was observed due to the release of most of the constituent water molecules. The outward OH flux and the concurrent diffusion flow in the grain structure might have caused a local accumulation of internal stress and thus produced hematite grain fracture. At this temperature, the shapes and sizes of the hematite grains were more or less remained the same like the original goethite. However, the GPMs exposed to 900°C showed the disappearance of hematite peak and appearance of the crystalline nepheline (sodium aluminium silicate, $AlNaSiO_4$) peaks even if the quartz and mullite phase was the dominant ones (Figure 7.21c). Compared to other mortars, the mortar prepared with 70% of WCP and high volume of FA showed some stable peaks at high temperatures.

Figure 7.22 illustrates the SEM micrographs of GPMs prepared with high WCP contents. The influence of elevated temperature exposure (400°C, 700°C and 900°C) on the microstructure of GPMs prepared with 50% and 70% of WCP was assessed. The structures of the GPMs were increasingly converted into less dense networks with microcracks and bigger pores with the rise in temperatures. The SEM images of the GPMs prepared with 70% of WCP after the exposure to ambient temperatures (400°C, 700°C and 900°C) were obtained from a crushed section. The Fe microcracks on the GPMs surface was observed when subjected to high temperatures where unreacted particles of the WCP, FA and some spherical holes were also evidenced. It is known that WCP contains several hollow spheres and the partial dissolution of such spherical particles can generate highly dispersed tiny pores in the matrix [1]. The voids of hollow cavities left by the dissolved WCP and FA particles in the matrix were filled by such unreacted tiny WCP spheres. In opposition, the mortar prepared with 70% of WCP showed more stable surface at high temperatures than the one made from 50% of the WCP.

7.9 SUMMARY

This chapter reported the effects of aggressive environmental conditions including the sulphate attack, sulphuric acid attack and elevated temperatures on the bond strength of GPMs containing various levels of POFA or WCP, GBFS and FA. The experiments conducted to assess these parameters were CS and bond tests performed in two cycles of 6 and 12 months for the sulphate and sulphuric acid attack, elevated temperatures from 400°C to 900°C tests. The XRD patterns and SEM images of GPMs were also recorded to determine the changes in their structures under aggressive environmental conditions. The most important conclusions of the study are as follows:

  i. Achievement of the high durability performance of GPMs (containing a high amount of WCP and FA in place of GBFS) against the sulphate and sulphuric acid environments.

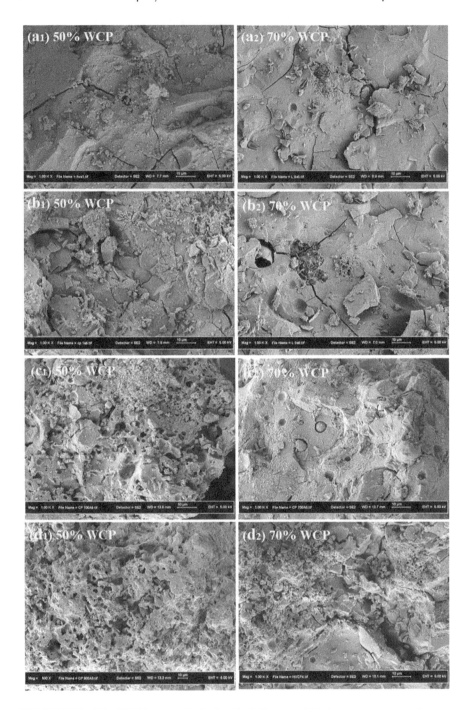

FIGURE 7.22 The SEM images displaying the influence of high temperatures on the morphologies of GPMs with different amounts of WCP: (a) 27°C, (b) 400°C, (c) 700°C, and (d) 900°C.

ii. Replacement of GBFS by WCP showed remarkable effect to reduce the loss in the bond strength between the GPMs and concrete substrate exposed to elevated temperatures up to 900°C.

iii. In each level of WCP-based GPMs, the substitution of FA in place of GBFS led to enhance the bond behaviour of specimens exposed to up to 900°C compared to other samples.

REFERENCES

1. Huseien, G.F., et al., Effects of POFA replaced with FA on durability properties of GBFS included alkali activated mortars. *Construction and Building Materials*, 2018. **175**: pp. 174–186.
2. Rashad, A.M., A comprehensive overview about the influence of different admixtures and additives on the properties of alkali-activated fly ash. *Materials & Design*, 2014. **53**: pp. 1005–1025.
3. Provis, J.L., A. Palomo, and C. Shi, Advances in understanding alkali-activated materials. *Cement and Concrete Research*, 2015. **78**: pp. 110–125.
4. Castel, A. and S.J. Foster, Bond strength between blended slag and Class F fly ash geopolymer concrete with steel reinforcement. *Cement and Concrete Research*, 2015. **72**: pp. 48–53.
5. Shekhawat, P., G. Sharma, and R.M. Singh, Strength behavior of alkaline activated eggshell powder and flyash geopolymer cured at ambient temperature. *Construction and Building Materials*, 2019. **223**: pp. 1112–1122.
6. Samadi, M., et al., Influence of glass silica waste nano powder on the mechanical and microstructure properties of alkali-activated mortars. *Nanomaterials*, 2020. **10**(2): p. 324.
7. Huseien, G.F., et al., Geopolymer mortars as sustainable repair material: a comprehensive review. *Renewable and Sustainable Energy Reviews*, 2017. **80**: pp. 54–74.
8. Du, K., C. Xie, and X. Ouyang, A comparison of carbon dioxide (CO_2) emission trends among provinces in China. *Renewable and Sustainable Energy Reviews*, 2017. **73**: pp. 19–25.
9. Huseien, G.F. and K.W. Shah, Durability and life cycle evaluation of self-compacting concrete containing fly ash as GBFS replacement with alkali activation. *Construction and Building Materials*, 2020. **235**: p. 117458.
10. Senthamarai, R. and P.D. Manoharan, Concrete with ceramic waste aggregate. *Cement and Concrete Composites*, 2005. **27**(9–10): pp. 910–913.
11. Huang, B., Q. Dong, and E.G. Burdette, Laboratory evaluation of incorporating waste ceramic materials into Portland cement and asphaltic concrete. *Construction and Building Materials*, 2009. **23**(12): pp. 3451–3456.
12. Senthamarai, R., P.D. Manoharan, and D. Gobinath, Concrete made from ceramic industry waste: durability properties. *Construction and Building Materials*, 2011. **25**(5): pp. 2413–2419.
13. Hussein, A.A., et al., Performance of nanoceramic powder on the chemical and physical properties of bitumen. *Construction and Building Materials*, 2017. **156**: pp. 496–505.
14. Samadi, M., et al., Properties of mortar containing ceramic powder waste as cement replacement. *Jurnal Teknologi*, 2015. **77**(12): pp. 93–97.
15. Huseien, G.F., et al., Waste ceramic powder incorporated alkali activated mortars exposed to elevated Temperatures: Performance evaluation. *Construction and Building Materials*, 2018. **187**: pp. 307–317.
16. Fernandes, M., A. Sousa, and A. Dias, *Environmental impacts and emissions trading-ceramic industry: a case study*. 2004: Technological centre of ceramics and glass, Portuguese association of ceramic industry (in Portuguese).

17. Pacheco-Torgal, F. and S. Jalali, Reusing ceramic wastes in concrete. *Construction and Building Materials*, 2010. **24**(5): pp. 832–838.
18. Mohammadhosseini, H., et al., Enhanced performance of green mortar comprising high volume of ceramic waste in aggressive environments. *Construction and Building Materials*, 2019. **212**: pp. 607–617.
19. Limbachiya, M., M.S. Meddah, and Y. Ouchagour, Use of recycled concrete aggregate in fly-ash concrete. *Construction and Building Materials*, 2012. **27**(1): pp. 439–449.
20. Heidari, A. and D. Tavakoli, A study of the mechanical properties of ground ceramic powder concrete incorporating nano-SiO_2 particles. *Construction and Building Materials*, 2013. **38**: pp. 255–264.
21. Huseien, G.F., et al., The effect of sodium hydroxide molarity and other parameters on water absorption of geopolymer mortars. *Indian Journal of Science and Technology*, 2016. **9**(48): pp. 1–7.
22. Huseien, G.F., et al., Effect of metakaolin replaced granulated blast furnace slag on fresh and early strength properties of geopolymer mortar. *Ain Shams Engineering Journal*, 2016. **9**(4): pp. 1557–1566.
23. Zhou, W., et al., A comparative study of high-and low-Al_2O_3 fly ash based-geopolymers: the role of mix proportion factors and curing temperature. *Materials & Design*, 2016. **95**: pp. 63–74.
24. Ranjbar, N., et al., Compressive strength and microstructural analysis of fly ash/palm oil fuel ash based geopolymer mortar. *Materials & Design*, 2014. **59**: pp. 532–539.
25. Rickard, W.D., et al., Assessing the suitability of three Australian fly ashes as an aluminosilicate source for geopolymers in high temperature applications. *Materials Science and Engineering: A*, 2011. **528**(9): pp. 3390–3397.
26. Chen, R., et al., Effect of particle size of fly ash on the properties of lightweight insulation materials. *Construction and Building Materials*, 2016. **123**: pp. 120–126.
27. Huseien, G.F., et al., Influence of different curing temperatures and alkali activators on properties of GBFS geopolymer mortars containing fly ash and palm-oil fuel ash. *Construction and Building Materials*, 2016. **125**: pp. 1229–1240.
28. Kumar, S., R. Kumar, and S. Mehrotra, Influence of granulated blast furnace slag on the reaction, structure and properties of fly ash based geopolymer. *Journal of Materials Science*, 2010. **45**(3): pp. 607–615.
29. Li, C., H. Sun, and L. Li, A review: The comparison between alkali-activated slag (Si+ Ca) and metakaolin (Si+ Al) cements. *Cement and Concrete Research*, 2010. **40**(9): pp. 1341–1349.
30. Deb, P.S., P. Nath, and P.K. Sarker, The effects of ground granulated blast-furnace slag blending with fly ash and activator content on the workability and strength properties of geopolymer concrete cured at ambient temperature. *Materials & Design (1980–2015)*, 2014. **62**: pp. 32–39.
31. Lee, N., E. Kim, and H. Lee, Mechanical properties and setting characteristics of geopolymer mortar using styrene-butadiene (SB) latex. *Construction and Building Materials*, 2016. **113**: pp. 264–272.
32. Yusuf, M.O., et al., Evolution of alkaline activated ground blast furnace slag–ultrafine palm oil fuel ash based concrete. *Materials & Design*, 2014. **55**: pp. 387–393.
33. Mirza, J., et al., Preferred test methods to select suitable surface repair materials in severe climates. *Construction and Building Materials*, 2014. **50**: pp. 692–698.
34. Alanazi, H., et al., Bond strength of PCC pavement repairs using metakaolin-based geopolymer mortar. *Cement and Concrete Composites*, 2016. **65**: pp. 75–82.
35. Mohammadhosseini, H., M.M. Tahir, and M. Sayyed, Strength and transport properties of concrete composites incorporating waste carpet fibres and palm oil fuel ash. *Journal of Building Engineering*, 2018. **20**: pp. 156–165.

36. Kubba, Z., et al., Impact of curing temperatures and alkaline activators on compressive strength and porosity of ternary blended geopolymer mortars. *Case Studies in Construction Materials*, 2018. **9**: p. e00205.
37. Mohammadhosseini, H., et al., Effects of waste ceramic as cement and fine aggregate on durability performance of sustainable mortar. *Arabian Journal for Science and Engineering*, 2019: pp. 1–12.
38. Lim, N.H.A.S., et al., The effects of high volume nano palm oil fuel ash on microstructure properties and hydration temperature of mortar. *Construction and Building Materials*, 2015. **93**: pp. 29–34.
39. Morgan, D., Compatibility of concrete repair materials and systems. *Construction and Building Materials*, 1996. **10**(1): pp. 57–67.
40. Geissert, D.G., et al., Splitting prism test method to evaluate concrete-to-concrete bond strength. *ACI Materials Journal*, 1999. **96**(3): pp. 359–366.
41. Momayez, A., et al., Comparison of methods for evaluating bond strength between concrete substrate and repair materials. *Cement and Concrete Research*, 2005. **35**(4): pp. 748–757.
42. Momayez, A, Comparison of methods for evaluating bond strength between concrete substrate and repair materials. *Cement and Concrete Research*, 2005**35**(4): pp. 748–757.
43. Fodil, D. and M. Mohamed, Compressive strength and corrosion evaluation of concretes containing pozzolana and perlite immersed in aggressive environments. *Construction and Building Materials*, 2018. **179**: pp. 25–34.
44. Concrete, A.C.C.-o. and C. Aggregates, *Standard specification for coal fly ash and raw or calcined natural pozzolan for use in concrete.* 2013: ASTM international.
45. Temuujin, J., A. van Riessen, and K. MacKenzie, Preparation and characterisation of fly ash based geopolymer mortars. *Construction and Building Materials*, 2010. **24**(10): pp. 1906–1910.
46. Gunasekara, C., et al., Zeta potential, gel formation and compressive strength of low calcium fly ash geopolymers. *Construction and Building Materials*, 2015. **95**: pp. 592–599.
47. Noushini, A. and A. Castel, The effect of heat-curing on transport properties of low-calcium fly ash-based geopolymer concrete. *Construction and Building Materials*, 2016. **112**: pp. 464–477.
48. ASTM, Standard test method for bond strength of epoxy-resin systems used with concrete by slant shear. *ASTM International*, 2005: pp. 1–8.
49. ASTM. *Standard test methods for fire tests of building construction and materials.* 2012: ASTM Philadelphia: pp. 1-10.
50. Standardization, I.O. *Fire-resistance tests: elements of building construction. commentary on test method and test data application.* 1994: ISO.
51. Ranjbar, N., et al., Compressive strength and microstructural analysis of fly ash/palm oil fuel ash based geopolymer mortar under elevated temperatures. *Construction and Building Materials*, 2014. **65**: pp. 114–121.
52. Ariffin, M., et al. Mix design and compressive strength of geopolymer concrete containing blended ash from agro-industrial wastes. in *Advanced Materials Research*. 2011. Trans Tech Publ. **339**:pp. 452–457.
53. Huseien, G.F., et al., Synergism between palm oil fuel ash and slag: Production of environmental-friendly alkali activated mortars with enhanced properties. *Construction and Building Materials*, 2018. **170**: pp. 235–244.
54. Huseien, G.F., J. Mirza, and M. Ismail, Effects of high volume ceramic binders on flexural strength of self-compacting geopolymer concrete. *Advanced Science Letters*, 2018. **24**(6): pp. 4097–4101.
55. Rashad, A.M., Properties of alkali-activated fly ash concrete blended with slag. *Iranian Journal of Materials Science and Engineering*, 2013. **10**(1): pp. 57–64.

56. Puertas, F., et al., Alkali-activated fly ash/slag cements: strength behaviour and hydration products. *Cement and Concrete Research*, 2000. **30**(10): pp. 1625–1632.
57. Dombrowski, K., A. Buchwald, and M. Weil, The influence of calcium content on the structure and thermal performance of fly ash based geopolymers. *Journal of Materials Science*, 2007. **42**(9): pp. 3033–3043.
58. Huseien, G.F. and K.W. Shah, Performance evaluation of alkali-activated mortars containing industrial wastes as surface repair materials. *Journal of Building Engineering*, 2020. **30**: pp. 101234.
59. Pacheco-Torgal, F., J. Castro-Gomes, and S. Jalali, Adhesion characterization of tungsten mine waste geopolymeric binder. Influence of OPC concrete substrate surface treatment. *Construction and Building Materials*, 2008. **22**(3): pp. 154–161.
60. Chen, M.-C., K. Wang, and L. Xie, Deterioration mechanism of cementitious materials under acid rain attack. *Engineering Failure Analysis*, 2013. **27**: pp. 272–285.
61. Matschei, T., F. Bellmann, and J. Stark, Hydration behaviour of sulphate-activated slag cements. *Advances in Cement Research*, 2005. **17**(4): pp. 167–178.

8 Performance Evaluation of Geopolymer as Repair Materials Under Freeze–Thaw Cycles

8.1 INTRODUCTION

Ordinary Portland Cement (OPC) has been widely used as concrete binder in various building substances. Large scale manufacturing of OPC has been shown to cause serious pollution in the environment in terms of considerable amount of greenhouse gas emissions [1–3]. In recent years, OPC-mediated CO_2 emission, landfill problems related to industrial wastes and low durability of traditional concrete are forcing researchers to explore alternative concretes with high strength and environmentally friendly materials. Annually, millions of tonnes of cement that is used for construction works worldwide is responsible for the enormous increase of CO_2 pollution, exploits huge amount of natural resources and consumes outsized energy in product stages [4–6]. To surmount these problems, usage of abundant and cheap industrial wastes to produce durable and green concrete binder emerged as obvious choice in its own right.

Geopolymer mortars (GPMs) are emergent constructional binder with lower CO_2 footprint compared to the conventional Portland cement [7–11]. GPM is an inorganic polymer material based on aluminosilicates (ASs) and calcium (Ca). This is produced from pozzolanic compounds with alkaline activator solution composed of sodium hydroxide (NH) and sodium silicate (NS) [12,13]. Alkali-activated binders are environmentally friendly, wherein their production consumes a moderate amount of energy [14,15]. Diverse industrial solid wastes containing silica, aluminium (Al) and/or Ca such as fly ash (FA), palm oil fuel ash (POFA), metakaolin and ground blast furnace slag (GBFS) are exploited to produce alkali-activated mortar/concrete [16–18].

Some South-East Asian nations especially Malaysia, Thailand and Indonesia produce plentiful of POFA as a waste by-product of the palm oil industries (one of the most important agro industries). The so-called waste POFA is obtained from the burning of empty fruit bunches, fibres and shells as fuel to generate electricity for oil production [19]. An estimate revealed that the total solid waste generated by the industry in Malaysia alone is about 10 million tons per year [20,21]. POFA disposal that causes serious environmental pollution is detrimental unless inhibited. Interestingly, recent research on the use of silica-rich POFA as cementitious material opened new gateway for the development of sustainable construction materials [17,22,23]. Another most common resource material for the production of GPMs is FA because it is abundant and cheap on earth. On top of these, FA also contains amorphous form of ASs [24]. Malaysia consumes about 8 million tonnes of coal for power generation annually [25]. Thus, it is believed that the use of FA in the production of geopolymer and alkali-activated mortar/concrete could lead to sustainable development in the construction sector [26,27]. It is worthy to mention that GBFS is a waste material obtained by quenching molten iron slag (a by-product of iron and steel-making) from a blast furnace in water or steam. In this process, a glassy and granular product is achieved which is then dried and ground into a fine powder called GBFS [28]. The chemical composition of a slag varies considerably depending on the raw materials used in the iron production process [29]. Due to high content of CaO and SiO_2, GBFS displays both cementing and pozzolanic properties [30]. In the past, GBFS has been widely used in the construction industry to improve the durability and mechanical properties of the conventional concrete [31]. It was reported that [32,33] alkali-activated slag exhibited significant contribution in terms of workability, compressive strength and durability. This prompted further impetus towards the usage of GBFS in making durable concretes.

To improve the mechanical and durable properties of GPM (binary blend) containing FA or POFA, various researchers [17,33,34] studied GBFS-incorporated concrete. This was simply because of their ability to generate secondary hydration that resulted in the formation of additional calcium–silicate–hydrate (C-S-H) [28,35,36]. Some researchers attempted to enhance the reactivity of FA in alkaline environment by increasing the GBFS content. The addition of calcium oxide (CaO) could form hydrated products such as C-S-H along with the ASs geopolymer network [37]. The amount of CaO content of the precursor materials was found to have considerable effect on the resulting hardened geopolymer [38,39]. Increase in the strength and decrease in the setting time was observed with the increase in CaO content [28]. However, the addition of CaO and calcium hydroxide [40] as a substitute of FA was found to improve the mechanical properties of the ambient cured samples. In short, this new inorganic environmentally friendly alkali-activated binder (free of Portland cement) with enhanced properties including high early strength, durability against chemical attack, high surface hardness and higher fire resistance became beneficial for the construction purposes. Among the pozzolanic materials, FA and GBFS have been extensively used for GPM production to enhance the durability properties. The distinctive advantages of GPMs or concrete have led the researchers to explore new types of binder materials. In this regard, POFA or WCP, FA and GBFS blended (a ternary blend) mortar activated with low concentration of alkaline activator solution still needs thorough investigation in terms of understanding the mechanisms of geopolymerization and associated improved attributes.

This chapter reports the influence of GBFS replacement by FA, POFA or WCP on the durability properties of GPMs (a ternary blend). Such mixes were prepared at varying concentrations of binder with alkaline solution activation to examine the feasibility of recycling industrial solid wastes and turning them into environmentally friendly, durable and sustainable binders/mortars. In this process, GBFS was replaced by various levels of FA and POFA or WCP in the practicable working range with suitable physical conditions necessary to produce these geopolymer ternary POFA or WCP, FA and GBFS mortar. All the designed mixes were characterized to evaluate the durability performance. Results were analysed, discussed and compared to achieve the optimum composition.

8.2 MIX DESIGN

Ternary binder contents (GBFS, FA and WCP or POFA) of varying proportions were used to prepare GPMs. Fifteen levels of replacement were adopted to evaluate the effect of WCP, FA and POFA contents as GBFS replacement on durability performance of proposed geopolymers. In each level, the minimum content of GBFS was kept to 20% as presented in Table 8.1. Furthermore, the respective values of NH molarity, NS to NH, alkaline solution to binder and binder to fine aggregate were selected to be 4 M, 0.75, 0.40 and 1.0, respectively, which were fixed for all mixes. Furthermore, at every level, the FA was replaced by POFA to evaluate the effect of increasing silica oxide and reducing aluminium oxide on GPMs properties. The content of SiO_2 was increased with the increase in POFA content. However, the content of Al_2O_3 was decreased with the increase in POFA content and reduction in FA content.

TABLE 8.1
The Compositions and Proportions of Various Components Used for the GPM Synthesis

Mix	Binder, Weight %				Fine Aggregates	Alkaline Solution	
	GBFS	FA	WCP	POFA		NaOH	Na_2SiO_3
$GPMs_1$	50	0	50	0	1.0	0.228	0.171
$GPMs_2$	40	10	50	0	1.0	0.228	0.171
$GPMs_3$	30	20	50	0	1.0	0.228	0.171
$GPMs_4$	20	30	50	0	1.0	0.228	0.171
$GPMs_5$	40	0	60	0	1.0	0.228	0.171
$GPMs_6$	30	10	60	0	1.0	0.228	0.171
$GPMs_7$	20	20	60	0	1.0	0.228	0.171
$GPMs_8$	30	0	70	0	1.0	0.228	0.171
$GPMs_9$	20	10	70	0	1.0	0.228	0.171
$GPMs_{10}$	50	50	0	0	1.0	0.228	0.171
$GPMs_{11}$	50	40	0	10	1.0	0.228	0.171
$GPMs_{12}$	50	30	0	20	1.0	0.228	0.171
$GPMs_{13}$	50	20	0	30	1.0	0.228	0.171
$GPMs_{14}$	50	10	0	40	1.0	0.228	0.171
$GPMs_{15}$	50	0	0	50	1.0	0.228	0.171

8.3 POROSITY

The porosity of repair material has a significant effect on durability. Generally, the dense, impermeable, highly resistive or non-conductive repair materials reveal the tendency where the repaired damaged area appears isolated from the adjacent undamaged areas. Consequently, the patched area of concrete shows a large difference in the porosity or chloride content from the rest, causing corrosion to remain localized in a limited region. The steel decay rate could accelerate, leading to early failure either in the scrap or the adjacent concrete. Thus, it is important to make sure that both concrete substrate and repair component have comparable porosity or density during the selection of a repair material. Keeping this in view, the porosity test was performed on sufficiently cured (at age of 28 days) cubic samples of dimension (50 mm × 50 mm × 50 mm) according to ASTM C642 standard.

Figure 8.1 depicts different FA-to-GBFS ratios on the porosity of GPMs at 28 days of age. The results revealed that the porosity decreased from 9.8 to 7.1 with corresponding increasing content of GBFS from 30% to 70%. The GPMs containing high level of FA showed enhanced water absorption and the increasing FA content led to reduce the dense gel formation with more pore structures. In short, an increase in GBFS level led to the formation of denser C-(A)-S-H gel, which in turn developed more homogeneous strength due to low porosity and water absorption.

Figure 8.2 displays the influence of POFA content on the porosity level of designed GPMs. Direct proportionality was observed between porosity values and POFA content. The porosity values were increased, respectively, from 13.3% to 19.1% when the POFA level was increased from 10% to 50%. This increase in the porosity with the increase in POFA content was due to the restricted dense gel formation. Ranjbar et al. [16] reported that an increase in silicate-to-aluminium ratio produced a negative effect on the compressive strength of GPM. Moreover, a reduction in alumina (Al) content with increasing POFA content showed a significant influence on the C-A-S-H product with lower strength and slower chemical reaction rates. According to Ariffin

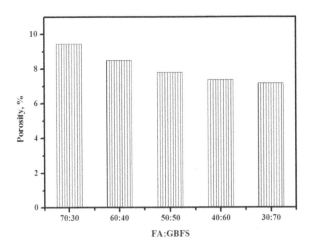

FIGURE 8.1 Effect of changing FA-to-GBFS proportions on the porosity of GPMs.

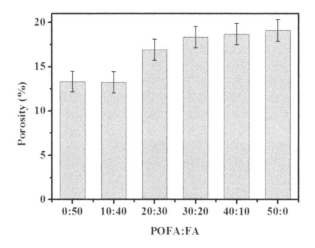

FIGURE 8.2 Effect of POFA:FA contents on porosity of GPMs.

et al. [41], low content of Al_2O_3 could be responsible for the reduced compressive strength. This observed reduction in the compressive strength was attributed to the incomplete geopolymerization process where Al_2O_3 revealed higher rate of dissolution during the early stage of geopolymerization. It was reported that the compressive strength and microstructure of GPMs could majorly depend on the silicate-to-aluminium ratio [42,43]. The ratio of Si:Al above 3.50 was found to negatively affect the strength and microstructure of GPMs [44].

Figure 8.3 shows the effect of high volume of WCP content and GBFS replaced by FA on the porosity of the studied GPMs. The porosity values were directly proportional to WCP content, where an increase in WCP in place of GBFS from 50% to

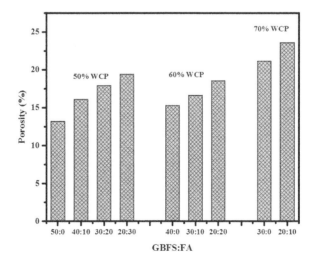

FIGURE 8.3 Effects of the WCP content on the porosity of the GPMs.

70% has led to an increase in the porosity of the GPMs from 13.2% to 21.1%, respectively. For 50%, 60% and 70% of WCP levels, the influence of FA replacement for GBFS on the porosity and water absorption of the GPMs was examined. The porosity values of GPMs were increased from 13.3% to 19.4% with an increase in the content of FA replacement for GBFS from 0% to 30%, respectively, with 50% WCP. Similar trend was observed with 60% and 70% of WCP containing GPMs. In general, as the FA content was increased and GBFS content was decreased, the porosity level of the GPMs was increased. The increasing levels of the WCP and FA caused an increase in the non-reacted and partially reacted particles, thereby reducing the C-S-H gel products with highly porous structure [45,46].

8.4 SURFACE ABRASION RESISTANCE

Abrasion resistance (AR) test of all GPMs was conducted at curing ages of 1, 3, 7 and 28 days under dry conditions following the specified Indian Standard (IS 1237–1980) whereby each specimen was weighed correctly by digital balance. After initial drying and weighing, specimens' thicknesses were measured at four different points. The grinding path of abrasion disc testing machine was evenly dispersed with 20 g abrasive powder (sand powder). GPMs were kept in the holding device of the abrasion machine, wherein a load of 300 N was subjected to the specimen. Next, the grinding machine was revolved at 30 rpm. The abrasive powder was constantly fed into the grinding trail to maintain a uniform track distribution related to the test specimen's width. Every specimen was abraded for 60 minutes from all sides and the reading was recorded after every 15 minutes. After the abrasion test, GPMs were weighed again to determine the weight loss. The thickness of each specimen was also recorded at four points. The degree of abrasion was estimated from the difference in the measured thickness before and after the testing.

Figure 8.4 presents the effects of varying FA-to-GBFS ratios on the grind depth of synthesized GPMs. The AR values (measured as grind depth) varied proportionately with the curing age and GBFS level. Proposed GPM specimens displayed higher AR

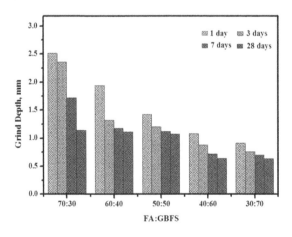

FIGURE 8.4 Effect of varying FA-to-GBFS ratios on the grind depth of GPMs.

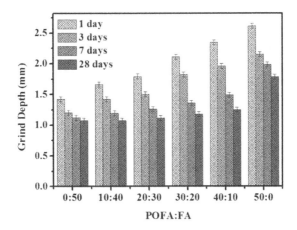

FIGURE 8.5 Effect of POFA:FA contents on AR of GPMs.

at 28 days of age than that of 1, 3 and 7 days. The AR was greatly influenced by GBFS content. The values of the grind depth of GPMs at 28 days of age decreased from 1.1 to 0.63 mm with the increasing GBFS content from 30% to 70%, correspondingly. With the increase in GBFS content, both the strength and AR were enhanced and the pores dropped. The AR was found to be directly proportional to the strength and inversely proportional to the porosity [47]. According to Liu et al. [48], the concrete with low porosity, high strength and strong interfacial bond could enhance the overall concrete abrasion–erosion resistance performance. As the porosity of the concrete decreases, the concrete becomes more impermeable and increases the AR of the GPMs. The results of compressive strength and water absorption were supported by the AR data. Wang et al. [49] demonstrated the reduction in compressive strength where an increase in the pore volume in hardened alkali-activated specimens have negatively affected the AR.

Figure 8.5 demonstrates the effect of POFA content changes on the AR of GPMs at different curing ages. The AR (high depth) was directly proportional to the age and POFA content. All the samples showed higher AR at 28 days of age compared to curing periods. Furthermore, the grind depth was increased respectively from 1.07 to 1.72 with increasing POFA content from 0% to 50% at 28 days of age. The AR of the GPM containing 10% of POFA, 40% of FA and 50% of GBFS was the highest (1.07 mm) at 28 days of age. The strength of the mortars was decreased with the increase in POFA content above 10%. This in turn enhanced the porosity and reduced the AR of the studied GPMs. The AR was directly proportional to the strength value and porosity of the mortars.

8.5 FREEZING–THAWING CYCLE RESISTANCE

The freeze–thaw cycling resistance test was carried out using prism-shaped GPM specimens of length 120 mm and cross-sectional area of 40 mm × 400 mm following the ASTM C666 standard (−17°C to 5°C). Another cubical specimen of dimension (50 mm × 50 mm × 50 mm) was also tested at curing age of 28 days. Method A was employed due to its more convenience than Method B. After curing for 28 days, all GPMs were subjected to thaw at room temperature to obtain the pulse velocity (UPV) and mass. Later, these GPMs were put in a container and submerged inside water, where the water temperature was controlled automatically by timer to attain every freeze–thaw cycle (a total of 300 cycles). The effectiveness of GPMs was evaluated depending on qualitative examination, loss in weight, UPV and residual compressive strength (RCS) development. Besides, the variation in the length and the dynamic modulus of GPMs were monitored after each 50 freeze–thaw cycles for a total of 300 cycles. The durability feature and the change in ultimate length were estimated at the end of freeze–thaw cycling using the relation:

$$Df = \frac{PN}{M} \tag{8.1}$$

where Df is durability factor of the test specimen, P is the relative dynamic modulus of elasticity at N cycles, N is number cycles at which P reached the specified minimum value for discontinuing the test and M is the specified number of cycles at which the exposure was terminated.

The bond strength between the normal concrete (NC) and GPMs was examined according to the ASTM C666 (Method A) wherein every specimen was subjected to a minimum of 300 freeze–thaw cycles to evaluate the slant shear bond strengths. It is important to note that by reducing the temperature of the mixes from 5°C to –20°C (75% of the cycle time) and increasing it from –20°C to 5°C (25% of the cycle time) within 5 hours for every cycle, it is possible to reach at the same situation of the freeze–thaw cycles. The durability test can be regarded as the most significant one for the prepared GPMs because no standard method was obtainable for the alkali-activated geopolymers. Therefore, it was implemented here for the studied temperature domain, temperature ramp and duration of cycles. Two approaches were proposed in the ASTM C666 standard including the Methods A and B. Method A comprised the freeze and thaw of the specimen in the water. In contrast, the Method B was composed of the freeze of the specimen in the air and thaw in the water. In the present work, the Method A was used due to its more convenience compared to Method B. The freezing was conducted following the approach A where nine mixes (composites of the NC and GPMs) were made in the cylindrical shape (of dimension 100 mm × 200 mm). At the 28 days of age, these samples were thawed at room temperature to record the slant shear bond strength. Later, these specimens were enclosed in the container to submerge in the water. The water temperature was controlled automatically using a timer to monitor every freeze–thaw cycle. After, these mixes were subjected to 300 freeze–thaw cycles, the durability performance of every GPM was evaluated after every 50 cycles based on the residual bond strength.

Failure of repair regions on the highways and bridges are often originated from the corrosion of concrete substrate and repair components. This failure can majorly be attributed to the exposure of the repair materials to the freeze–thaw cycling. It is worth to examine the FA-to-GBFS ratio–dependent RCS, internal frost damage and surface scaling of GPMs at different freeze–thaw cycles. All the samples were exposed to 0–300 freeze–thaw cycles on GPMs cured at 28 days of age and the RCS

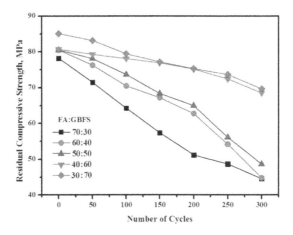

FIGURE 8.6 Freeze–thaw cycle–dependent RCS of proposed geopolymers.

was measured (Figure 8.6). The influence of varying GBFS levels (30%–70%) on the durability of GPMs exposed to freeze–thaw cycles was evaluated in terms of RCS. The value of RCS increased from 44.2 to 69.6 MPa with increasing level of GBFS from 30% to 70%, correspondingly. Internal frost damage increased with the increase in number of freeze–thaw cycles and decreased with increasing GBFS content. Likewise, the remaining weight and surface scaling of the prepared GPMs decreased with the increase in GBFS content. The achieved high durability of GPMs at elevated GBFS level can clearly be observed from the RCS, residual weight and surface scaling (deterioration) results. Additionally, the porosity reduced and the pores refined with the increase in GBFS content, thereby contributing to the reduction in the frost formation. Meanwhile, the pores became more disconnected, leading to the reduction in the capillary transport of external liquid into the concrete pores during freeze–thaw cycles. This in turn created less frost growth, which was supposed to be a major scaling mechanism governed by cryogenic suction of surface liquid under freezing.

Table 8.2 depicts the durability factor and length change of the studied GPMs under varying freeze–thaw cycles (average of three specimens). Higher durability factor indicates that the material is better to be used when exposed to severe

TABLE 8.2
Freeze–Thaw Cycle–Dependent Durability of the Prepared GPMs

Alkali-Activated Mortars	Number of Cycles	Durability Factor (±1)	Length Change (%)
GPM_1	300	79	−0.19
GPM_2	300	87	−0.14
GPM_3	300	91	−0.08
GPM_4	300	94	−0.05
GPM_5	300	99	−0.02

cold condition. For instance, alkali-activated repair material containing 50% GBFS showed a durability factor over 90%. Conversely, the repair material prepared with 30% GBFS revealed low durability factor (79%), which is not appropriate for the place (environment) exposed to severe cold.

Figures 8.7 and 8.8 illustrate the effect of POFA replacement on the residual strength and weight loss of the studied GPMs, respectively. The loss of strength and weight was directly proportional to the POFA content. The strength of the mortar dropped to 20.1 MPa when the POFA content increased to 50%. This in turn led to a weight loss higher than 5% after 300 freezing–thawing cycles compared to 48.6 MPa and weight loss of 98% with 0% POFA. With the increasing freezing–thawing cycles and POFA content, the damages occurred mainly on the surfaces and edges of the specimens (Figure 8.9; marked by red squares and circles). Furthermore, the deterioration was the highest for GPM containing 50% of POFA.

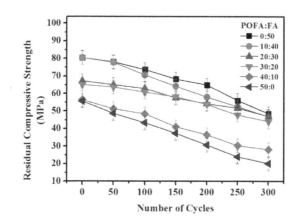

FIGURE 8.7 Effect of POFA replacement by FA on the RCS of GPMs.

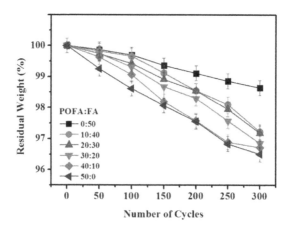

FIGURE 8.8 Effect of POFA replacement by FA on the residual weight of GPMs.

Evaluation of GP Under Freeze–Thaw Cycles

FIGURE 8.9 Effect of POFA replacement by FA on the surface textures of GPMs.

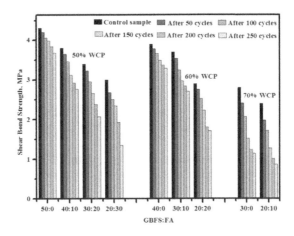

FIGURE 8.10 Effect of acid attack on shear bond strength of GPMs.

The influence of freeze–thaw cycles on the bond strength performance of the GPM specimens is illustrated in Figure 8.10. The tested specimens showed an inverse relationship between the residual bond strength and freeze–thaw cycles, and the WCP and FA contents. The rise in the content of WCP from 50% to 70% caused an increase in the loss percentage on the bond strength from 13.8% to 58.4% after 250 freeze–thaw cycles, respectively. Similar trend was observed for the residual bond strength at high amount of FA as GBFS replacement in each level of WCP matrix where the loss percentage on the bond strength was increased from 26.7% to 54.3% with the increase in the FA content from 10% to 30% compared to 13.8% with FA content of 0%. It was found that the increase in the content of WCP and FA could enhance the amount of non-reacted and partly reacted silicate and minimized the

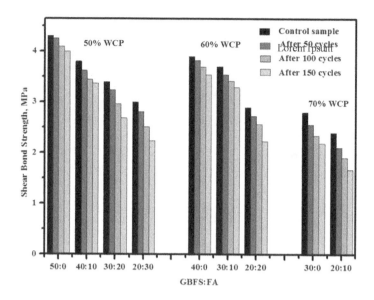

FIGURE 8.11 Effect of the acid attack on the shear bond strength of the GPMs.

final C-S-H product. The amount of the porous structure in the prepared specimens was increased with the increase in WCP and FA where high quantity of non-reacted and partly reacted silicate was formed. The existence of the voids aided the intensification of the ice and destroyed the particles interlocking [50]. This was attributed to the reduction of the GPMs resistance to the freeze–thaw cycles, indicating higher loss in the bond strength.

Figure 8.11 illustrates the residual bond strength of the GPMs enclosing high level of WCP plus GBFS and FA when exposed to the wet–dry cycles at 28 days of age. For all the GPMs, a direct relation was observed between the wet–dry cycle numbers and loss in the bond strength. The results showed a significant effect of the high volume of WCP and FA on the strength loss of GPM specimens, the percentage of the total loss in the bond strength was increased from 6.9% to 21.7% with the increase in the WCP level from 50% to 70% replacement for the GBFS, respectively. Also, the strength loss was influenced by the level of FA replaced GBFS in the high-volume WCP mortar specimens. Increase in the FA level from 0% to 30% negatively affected the microstructures of the specimens and led to an increase in the strength loss from 6.9% to 25.3%, respectively. According to the previous findings [51], the main reason for the loss of the strength was due to the increase in overall porosity with the rise in WCP and FA contents, where no pores were available in the matrix. This greatly favoured the water entry into the matrix during the wet–dry cycles. The internal and external deterioration was increased, leading to a loss in the bond strength and low durability over time.

8.6 DRYING SHRINKAGE

The drying shrinkage test was performed based on the procedure outlined in ASTM C157/C157M. Three sets of GPMs (three prisms for each mixture), each of dimension

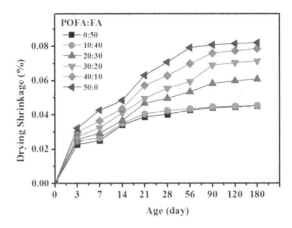

FIGURE 8.12 Curing age–dependent variation of drying shrinkage of GPMs.

25 mm × 25 mm × 250 mm in the form of prism, were used. These specimens were prepared in accordance with ASTM C192/192M and cured in the ambient condition. Stainless steel studs were embedded into the specimens to facilitate the measurement of length change. Next, these mortar specimens were de-moulded 24 hours after casting before being moved to a constant temperature environment maintained at 23°C ±1°C and relative humidity of 50%. Subsequently, the readings were taken using demec meter at 1–14, 21, 28, 56, 90 and 180 days.

Figure 8.12 illustrates the effect of POFA replacement on the drying shrinkages of GPMs for different curing ages. For all GPMs samples, the shrinkage values increased with the increasing age. The drying shrinkage values were directly proportional to the POFA content and curing ages. The drying shrinkage of the GPM cured for 180 days varied in the range of 450–820 when the POFA level was 50%. The values of dry shrinkage were increased from 420 to 818 macrostrains with increasing POFA content from 0% and 50%, respectively. This observed improvement in the drying shrinkage of the GPMs was attributed to the presence of less interconnected capillary network of the alkali-activated matrix, which was consistent with the report of Deb et al. [52].

8.7 WET–DRY CYCLE RESISTANCE

No standard method for the wet–dry cycle durability test was available. In the tropical nation like Malaysia, the weather condition is random and vary fast from hot to dry (for few days) to rainy. Thus, those few days were very significant during the conduction of such test. This test was designed to mimic the natural Malaysian environmental condition for accelerating the wet–dry cycles (Table 8.3). Figure 8.13 shows the condition for the wet–dry test cycles wherein the data in each 50 cycles were recorded to determine any alteration in the loss of shear bond strength. The total wet–dry cycles were adopted to evaluate the durability of the specimens.

Figure 8.14 depicts the influence of POFA:FA changes on wet–dry cycles resistance of GPMs. The strength loss of GPMs was directly correlated to the POFA

TABLE 8.3
Test Conditions for Cyclic Wetting and Drying

Specimen size	50 mm × 50 mm × 50 mm
Specimens number	Three specimens for each GPM mixture
Wet condition	27°C
Dry condition	65°C
Cycle	3 days dry condition and 1 day wet condition
Total cycles	150 cycles

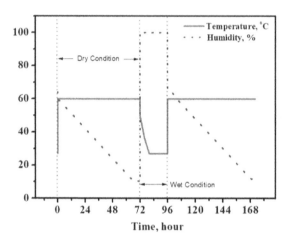

FIGURE 8.13 The GPM's wet–dry cyclic process for one cycle.

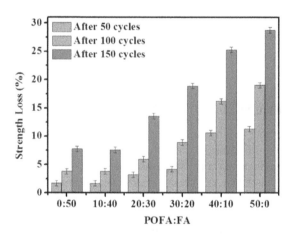

FIGURE 8.14 Effect of different wet–dry cycles on strength loss of GPMs containing POFA:FA in various ratios.

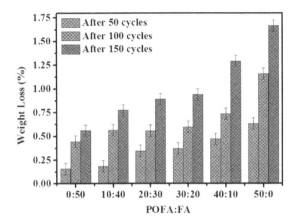

FIGURE 8.15 Effect of POFA on wet–dry cycle resistance (weight loss) of GPMs.

content. The strength loss values of the GPMs were increased from 7.7% to 27.8% with increased POFA content from 0% to 50%. The weight loss was influenced considerably by the POFA content variation. For 10% of POFA at 150 cycles, the GPM revealed low strength loss (7.5%). However, the strength loss percentage was increased to 28.7% with an increase in POFA level up to 50%.

Figure 8.15 displays the POFA content–dependent weight loss of GPMs after 150 wet–dry cycles. The weight loss values increased from 0.5% to 1.6% with the increase in POFA content from 0% to 50%, respectively. The dense structure and low porosity of GPM at 10% of POFA could reduce the internal and external damage and enhance the resistance to wet–dry cycles compared to the specimens containing higher than 10% POFA levels.

8.8 SUMMARY

For the first time, we investigated the influence of increasing FA and POFA or WCP replacement in the proposed ternary blends containing various levels of GBFS. The main objective was to produce environmental friendly, cheap and robust binder materials from industrial and agriculture wastes effective for construction purposes. The durability properties of these new GPMs were evaluated in terms of porosity, resistance for dry shrinkage and freezing–thawing cycles, abrasion and wet–dry cycles. Based on the experimental results, the following conclusions were drawn:

i. GBFS-incorporated GPMs were prepared by replacing FA with POFA.
ii. The GPM porosity is highly influenced by POFA, FA and WCP contents. It was found that with increasing high-content silica materials and with decreasing calcium oxide content, the porosity tends to drop.
iii. Inclusion waste materials such as POFA, FA and WCP improve the suggested mortar performance as repair materials by reducing the total drying shrinkage.

iv. GPMs containing high contents of WCP or POFA showed poor performance when exposed to 150 or higher freezing–thawing and wet–dry cycles.
v. Durability of GPMs revealed strong sensitivity to the POFA content variation.
vi. 10% of POFA replacing FA achieved optimum AR of GPMs.
vii. Freezing–thawing resistance of GPMs was inversely correlated to POFA content.

REFERENCES

1. Duxson, P., et al., Geopolymer technology: the current state of the art. *Journal of Materials Science*, 2007. **42**(9): p. 2917–2933.
2. Rashad, A.M., A comprehensive overview about the influence of different admixtures and additives on the properties of alkali-activated fly ash. *Materials & Design*, 2014. **53**: p. 1005–1025.
3. Huseien, G.F., et al., Geopolymer mortars as sustainable repair material: a comprehensive review. *Renewable and Sustainable Energy Reviews*, 2017. **80**: p. 54–74.
4. Zhang, Y., et al., Aspen plus-based simulation of a cement calciner and optimization analysis of air pollutants emission. *Clean Technologies and Environmental Policy*, 2011. **13**(3): p. 459–468.
5. Huseien, G.F., et al., Synthesis and characterization of self-healing mortar with modified strength. *Jurnal Teknologi*, 2015. **76**(1): p. 195–200.
6. Hussein, A.A., et al., Performance of nanoceramic powder on the chemical and physical properties of bitumen. *Construction and Building Materials*, 2017. **156**: p. 496–505.
7. Pacheco-Torgal, F., et al., *Alkali-activated cement-based binders (AACBs) as durable and cost-competitive low-CO_2 binder materials: some shortcomings that need to be addressed.* 2017: Butterworth-Heinemann.
8. Provis, J.L., A. Palomo, and C. Shi, Advances in understanding alkali-activated materials. *Cement and Concrete Research*, 2015. **78**: p. 110–125.
9. Turner, L.K. and F.G. Collins, Carbon dioxide equivalent (CO_2-e) emissions: a comparison between geopolymer and OPC cement concrete. *Construction and Building Materials*, 2013. **43**: p. 125–130.
10. Habert, G. and C. Ouellet-Plamondon, Recent update on the environmental impact of geopolymers. *RILEM Technical Letters*, 2016. **1**: p. 17–23.
11. Ouellet-Plamondon, C. and G. Habert, Life cycle assessment (LCA) of alkali-activated cements and concretes, in *Handbook of alkali-activated cements, mortars and concretes*. 2015, Elsevier. p. 663–686.
12. Karakoç, M.B., et al., Mechanical properties and setting time of ferrochrome slag based geopolymer paste and mortar. *Construction and Building Materials*, 2014. **72**: p. 283–292.
13. Huseien, G.F., et al., Effect of metakaolin replaced granulated blast furnace slag on fresh and early strength properties of geopolymer mortar. *Ain Shams Engineering Journal*, 2016. **9**(4): 1557–1566.
14. Attanasio, A., et al., Alkali-activated mortars for sustainable building solutions: effect of binder composition on technical performance. *Environments*, 2018. **5**(3): p. 35.
15. Shi, C., A.F. Jiménez, and A. Palomo, New cements for the 21st century: the pursuit of an alternative to Portland cement. *Cement and Concrete Research*, 2011. **41**(7): p. 750–763.
16. Ranjbar, N., et al., Compressive strength and microstructural analysis of fly ash/palm oil fuel ash based geopolymer mortar under elevated temperatures. *Construction and Building Materials*, 2014. **65**: p. 114–121.

17. Salih, M.A., et al., Development of high strength alkali activated binder using palm oil fuel ash and GGBS at ambient temperature. *Construction and Building Materials*, 2015. **93**: p. 289–300.
18. Huseiena, G.F., et al., *Potential use coconut milk as alternative to alkali solution for geopolymer production*. 2016. **78**(11): p. 133–139.
19. Yusuf, T.O., et al., Impact of blending on strength distribution of ambient cured metakaolin and palm oil fuel ash based geopolymer mortar. *Advances in Civil Engineering*, 2014. **2014**: p. 1–9.
20. Ismail, M., et al. Early strength characteristics of palm oil fuel ash and metakaolin blended geopolymer mortar. in *Advanced materials research*. 2013. Trans Tech Publ. **690**: p. 1045–1048.
21. Islam, A., et al., The development of compressive strength of ground granulated blast furnace slag-palm oil fuel ash-fly ash based geopolymer mortar. *Materials & Design*, 2014. **56**: p. 833–841.
22. Khankhaje, E., et al., Sustainable clean pervious concrete pavement production incorporating palm oil fuel ash as cement replacement. *Journal of Cleaner Production*, 2017. **172**: 1476–1485.
23. Khankhaje, E., et al., On blended cement and geopolymer concretes containing palm oil fuel ash. *Materials & Design*, 2016. **89**: p. 385–398.
24. Zhou, W., et al., A comparative study of high-and low-Al_2O_3 fly ash based-geopolymers: the role of mix proportion factors and curing temperature. *Materials & Design*, 2016. **95**: p. 63–74.
25. Ranjbar, N., et al., Compressive strength and microstructural analysis of fly ash/palm oil fuel ash based geopolymer mortar. *Materials & Design*, 2014. **59**: p. 532–539.
26. Rickard, W.D., et al., Assessing the suitability of three Australian fly ashes as an aluminosilicate source for geopolymers in high temperature applications. *Materials Science and Engineering: A*, 2011. **528**(9): p. 3390–3397.
27. Chen, R., et al., Effect of particle size of fly ash on the properties of lightweight insulation materials. *Construction and Building Materials*, 2016. **123**: p. 120–126.
28. Huseien, G.F., et al., Influence of different curing temperatures and alkali activators on properties of GBFS geopolymer mortars containing fly ash and palm-oil fuel ash. *Construction and Building Materials*, 2016. **125**: p. 1229–1240.
29. Kumar, S., R. Kumar, and S. Mehrotra, Influence of granulated blast furnace slag on the reaction, structure and properties of fly ash based geopolymer. *Journal of Materials Science*, 2010. **45**(3): p. 607–615.
30. Li, C., H. Sun, and L. Li, A review: the comparison between alkali-activated slag (Si+Ca) and metakaolin (Si+Al) cements. *Cement and Concrete Research*, 2010. **40**(9): p. 1341–1349.
31. Deb, P.S., P. Nath, and P.K. Sarker, The effects of ground granulated blast-furnace slag blending with fly ash and activator content on the workability and strength properties of geopolymer concrete cured at ambient temperature. *Materials & Design (1980–2015)*, 2014. **62**: p. 32–39.
32. Lee, N., E. Kim, and H. Lee, Mechanical properties and setting characteristics of geopolymer mortar using styrene-butadiene (SB) latex. *Construction and Building Materials*, 2016. **113**: p. 264–272.
33. Yusuf, M.O., et al., Evolution of alkaline activated ground blast furnace slag–ultrafine palm oil fuel ash based concrete. *Materials & Design*, 2014. **55**: p. 387–393.
34. Rashad, A.M., Properties of alkali-activated fly ash concrete blended with slag. *Iranian Journal of Materials Science and Engineering*, 2013. **10**(1): p. 57–64.
35. Phoo-ngernkham, T., et al., High calcium fly ash geopolymer mortar containing Portland cement for use as repair material. *Construction and Building Materials*, 2015. **98**: p. 482–488.

36. Nath, P., P.K. Sarker, and V.B. Rangan, Early age properties of low-calcium fly ash geopolymer concrete suitable for ambient curing. *Procedia Engineering*, 2015. **125**: p. 601–607.
37. Al-Majidi, M.H., et al., Development of geopolymer mortar under ambient temperature for in situ applications. *Construction and Building Materials*, 2016. **120**: p. 198–211.
38. Izquierdo, M., et al., Coal fly ash-slag-based geopolymers: microstructure and metal leaching. *Journal of Hazardous Materials*, 2009. **166**(1): p. 561–566.
39. Yip, C.K., G. Lukey, and J. Van Deventer, The coexistence of geopolymeric gel and calcium silicate hydrate at the early stage of alkaline activation. *Cement and Concrete Research*, 2005. **35**(9): p. 1688–1697.
40. Sugama, T., L. Brothers, and T. Van de Putte, Acid-resistant cements for geothermal wells: sodium silicate activated slag/fly ash blends. *Advances in Cement Research*, 2005. **17**(2): p. 65–75.
41. Ariffin, M., et al. Mix design and compressive strength of geopolymer concrete containing blended ash from agro-industrial wastes. in *Advanced materials research*. 2011: Trans Tech Publ. **339**: p. 452–457.
42. Duxson, P., et al., The role of inorganic polymer technology in the development of 'green concrete'. *Cement and Concrete Research*, 2007. **37**(12): p. 1590–1597.
43. Huseien, G.F., et al., Synergism between palm oil fuel ash and slag: Production of environmental-friendly alkali activated mortars with enhanced properties. *Construction and Building Materials*, 2018. **170**: p. 235–244.
44. Chindaprasirt, P., et al., Effect of SiO_2 and Al_2O_3 on the setting and hardening of high calcium fly ash-based geopolymer systems. *Journal of Materials Science*, 2012. **47**(12): p. 4876–4883.
45. Huseien, G.F., et al., Properties of ceramic tile waste based alkali-activated mortars incorporating GBFS and fly ash. *Construction and Building Materials*, 2019. **214**: p. 355–368.
46. Huseien, G.F., et al., Effects of ceramic tile powder waste on properties of self-compacted alkali-activated concrete. *Construction and Building Materials*, 2020. **236**: p. 117574.
47. Mohebi, R., K. Behfarnia, and M. Shojaei, Abrasion resistance of alkali-activated slag concrete designed by Taguchi method. *Construction and Building Materials*, 2015. **98**: p. 792–798.
48. Liu, Y.-W., T. Yen, and T.-H. Hsu, Abrasion erosion of concrete by water-borne sand. *Cement and Concrete Research*, 2006. **36**(10): p. 1814–1820.
49. Wang, S.-D., K.L. Scrivener, and P. Pratt, Factors affecting the strength of alkali-activated slag. *Cement and Concrete Research*, 1994. **24**(6): p. 1033–1043.
50. Cai, L., H. Wang, and Y. Fu, Freeze–thaw resistance of alkali–slag concrete based on response surface methodology. *Construction and Building Materials*, 2013. **49**: p. 70–76.
51. Chang, H., et al., Influence of pore structure and moisture distribution on chloride "maximum phenomenon" in surface layer of specimens exposed to cyclic drying-wetting condition. *Construction and Building Materials*, 2017. **131**: p. 16–30.
52. Deb, P.S., P. Nath, and P.K. Sarker, Drying shrinkage of slag blended fly ash geopolymer concrete cured at room temperature. *Procedia Engineering*, 2015. **125**: p. 594–600.

9 Methods of Evaluating the Geopolymer Efficiency as Alternative Concrete Surface Repair Materials Compared to Commercials Products

9.1 INTRODUCTION

Many infrastructures across the world exist in advanced state of degradation because of mechanical and physical factors such as surface degradation (abrasion, erosion, cavitation, impact and scaling), internal cracking (crystallization, permanent or excessive structural loading) and exposure to extreme temperatures (such as fire and freezing). Viaducts, parking and many other structures are threatened by unexpected collapse at any time without being able to predict the time of their collapse though the use of very sophisticated destructive and nondestructive apparatus. Thus, this results in heavy human losses and material damages. The loss of infrastructure also leads to painful economic damage. Thus, a partial repair that can shorten the duration of rehabilitation would be better than a total demolition and reconstruction.

In severe climatic conditions, the surfaces of concrete sidewalks, parking decks, bridges, canals, dams and other structures deteriorate progressively due to a variety of causes. For their repair and maintenance, countless surface repair mortars are abundantly available on the market but are constantly used before they have been tested in the laboratory [1]. In the last few years, many materials and methods have been developed to repair concrete. Sales representatives who are selling repair materials promise wondrous results with their products [2,3]. Information on these products has always been scarce and manufacturers have been unable to supply specific data on these mortars' resistance to harsh conditions found in many parts of the globe. Even if data are available, it is usually for room temperature conditions and is therefore of very little value for structures exposed to severe hot and cold climatic conditions. Some experts also estimate that up to half of all concrete repairs fail. Many of the materials do not work, and concrete repairs are tricky. There are few engineers who have adequate knowledge of concrete repairs, and contractors with experience in concrete repairs are scarce too.

The partial repair procedure can be processed to extend the lifespan of structure by ensuring at least the same mechanical and durability properties than those of the existing concrete. In this repair, a total demolition and the associated expensive economic consequences can be avoided.

The repair efficiency depends on the quality of the repair material and its capacity to fill the section to be repaired and to cover the reinforcement. The efficiency also varies with the repair method and the compatibility between the substrate and the repair material to ensure a long-term total bond between the two materials. In addition, the capacity of the repair material to completely fill the restrained spacing is the function of its capacity to flow, type of repair materials (concrete, mortar, etc.), width of the repair zone and steel reinforcement density in this zone [4]. It is important that every concrete structure should continue to perform its intended functions and maintain its required strength and serviceability, throughout the specified or expected service life. Recently, developments in cement and concrete technology have concentrated on achieving higher and higher strengths. Both strength and durability must explicitly be considered at the design stage [5].

Concrete repair is a complex process, and the current experiences with concrete repair are not satisfying. Repair materials are often perceived to lack both early age performance and long-term durability, due to the inherent brittleness and susceptibility to fracture.

Many undesirable repair behaviours were observed on the field in the forms of early age surface cracking or interface de-lamination between the repair and the concrete substrate, due to relative volume change of repair material and substrate concrete. Cracking and delamination are the common causes of many repair pathologies. They facilitate the ingress of chlorides, oxygen, moisture, alkali or sulphates into the repaired system and accelerate further deterioration. Furthermore, the loss of structural integrity due to the cracking or the de-laminating impairs load transfer between the repair and the concrete substrate. To make successful repairs with maximum life, it must have the system to select suitable repair materials without adventures.

The key to select an appropriate repair material is to understand its purpose in the repair. More often than not, many users in the repair industry believe that the simple answer to the repair problems is improving the compressive strength of the repair material or accelerating its strength gain to reduce disruption to the commuting public [6,7]. However, compressive strength is not an important material property for selecting a repair material as observed in the literature review. These demands have resulted in an emergence of a range of new rapid set of repair material products, not all of which perform equally or adequately.

In civil engineering structures, many causes can lead to degradation of concrete and cracks and these can occur relatively soon after the structure is built. Repair of conventional concrete structures usually involves applying a concrete and mortar which is bonded to the damaged concrete surface. Sometimes, the concrete and mortar need to be keyed into the existing structure with the metal pins to ensure that it does not fall away. Repairs can be particularly time-consuming and expensive because it is often very difficult to gain access to the structure to make repairs, especially if they are underground or at a great height. Repairs should be considered in terms of their cost and estimated service life to give a clear picture of which repair will truly be the most cost-effective. Understanding the specific conditions of which are necessary to increase repair, reliability and service life will create a basic knowledge to when repairs should be implemented.

As mentioned earlier, the main problems are as follows:

i. A large number of manufacturers.
ii. Hundreds of materials available or under development.
iii. Data either unavailable or available only for ambient temperatures.
iv. Very little data available on materials exposed to severe climatic conditions.
v. Similar products from different manufacturers can give different results.
vi. Increasingly difficult for practicing engineers to select the right and appropriate product for a given job.
vii. Specific need to select appropriate repair materials for severe climatic conditions.

9.2 CAUSES OF CONCRETE SURFACE DEGRADATION

The deterioration can be due to either external factors or the internal causes within the concrete itself under physical, chemical or mechanical actions. Mechanical damage is caused by impact, abrasion, cracking, erosion, cavitation or contraction. Chemical causes of deterioration include carbonation, alkali–silica, alkali–carbonate reactions and efflorescence. External chemical attack occurs mainly through the action of aggressive ions, such as chlorides, sulphates or carbon dioxide, and many natural or industrial liquids and gases [6].

Physical causes of deterioration include the effects of high temperature or of differences in thermal expansion of aggregate and of the hardened cement paste. The alternating freezing and thawing of concrete and the associated action of the de-icing salts are the important causes of deterioration. Physical and chemical processes of deterioration can act in a synergistic manner, such as the effect of sea water on concrete [8].

The two main causes of concrete degradation are as follows:

i. Surface degradation: abrasion, erosion, cavitations and scaling.
ii. Internal cracking: humidity or temperature gradient, crystallizing pressures, structural loading and exposure to extreme temperatures (freezing, fire).

The principal degradation origins in the following points:

i. Climatic and environmental factors including the changes in concrete structures according to climatic conditions and type of exposure to aggressive agents. At cold temperatures and high humidity, internal cracking due to freeze–thaw, spilling due to de-icing salts, reinforcement corrosion and alkali-aggregate reactions can be induced. At warm temperature and humid conditions, aggressive water attack and alkali-aggregate reactions can occur. At dry climatic conditions, carbonation phenomenon can appear. At

marine environment, sea water attack (sulphates), reinforcement. corrosion, glace abrasion and freeze–thaw deterioration can occur.
ii. Other origins related to structure design, structure placement and characteristics of concrete materials, and structure maintenance.

Thus, special care should be taken on the concrete placement methods, steel reinforcement placement, curing methods, formwork rigidity, design and sealing, excessive bleeding and plastic shrinkage. Other important parameters affecting the durability of concrete are characteristics of constituent materials (cement type, aggregate type and mineral additives), concrete compressive and splitting tensile strengths and water-to-binder ratio.

In severe climatic conditions, the surfaces of concrete deteriorate progressively due to variety of causes. For their repair and maintenance, countless surface repair mortars are abundantly available on the market and are constantly used without prior testing in the laboratory. For this reason, Mirza et al. [1] used 40 different mortars comprising cement-based mortars, polymer-modified cement-based mortars, containing styrene–butadiene rubber and acrylics, epoxy mortars and emulsified epoxy mortars from different manufacturers were subjected to a battery of mechanical and durability tests. These tests included bond strength, abrasion–erosion resistance, shrinkage–expansion, compressive strength, and thermal compatibility with base concrete. Test data are obtained from these tests. The test data revealed that over 65% (dry cure) and 89% (wet cure) of the mortars had a bond strength better than the reference cement mortar, while over 90% performed better in the abrasion–erosion resistance test. Similarly, over 80% of the mortars exhibited higher compressive strength (84% in dry curing and 81% in wet curing) than the reference mortar. In the shrinkage–expansion test, 53% and 66% of the surface repair mortars showed lower than 0.15% net change and 0.2% total change, respectively, as specified in ASTM C928. However, in the thermal compatibility with base concrete test, only 36% of the mortars performed better indicating its importance and preference in severe climatic conditions.

9.3 COMMERCIAL REPAIR MATERIALS

A wide variety of surface repair materials are now available to the manufactures, which can be classified into three primary groups: cementitious mortars, polymer-modified cementitious mortars and resinous mortars. The choice of suitable repair materials for use in reducing damage to concrete structure must be conducted with care. Before the selection of materials that would prove to be the most effective from among a wide variety of existing coating materials, one should first take into consideration the following characteristics:

i. Concrete is a porous material though which moisture evaporates constantly from the surface. The repairing materials must therefore be permeable and allow water vapour to evaporate freely from the concrete surface.

ii. There are huge number of commercial repairing materials available in the market. Therefore, to select the best materials suitable for the project requirement, consideration must be given not only to selecting the right material but also to the method of application to the damaged concrete surfaces.

The suitable repairing materials should have the following characteristics:

   i. Good bonding strength.
   ii. High abrasion resistance.
   iii. Strong resistance to freeze–thaw and wet–dry cycles.
   iv. Good permeability.
   v. Modulus of elasticity and coefficient of thermal expansion as possible to the base concrete.
   vi. Good adhesion in dry, damp and wet conditions.
   vii. Low shrinkage during curing, early age and long term.
   viii. Withstand ageing and weathering conditions.

In view of the earlier mentioned characteristics, the selection of repairing materials will be based on the following: (i) cement-based, (ii) polymer-modified cement-based and (iii) epoxy-based materials.

9.3.1 Cement-Based Materials

Two types of cement-based materials reviewed in this section included fibres reinforced and cement mortar containing silica fume. It was reported [9] that to achieve a better performance, a bonding material between the existing concrete and fresh mortar is often required. One of the best characteristics of repairing materials is that their porosity and permeability will be similar to that of the base concrete.

For fibre-reinforced cement, the micro reinforcement in the cement-based materials will be fibre mesh, fibre glass and stainless-steel strip. This mortar as commercial surface repair materials exhibits the following:

   i. High-impact resistance.
   ii. High ductility.
   iii. Good bonding.
   iv. Resistance to spalling, cracking and weathering.
   v. High compressive and tensile strength.
   vi. High abrasion resistance.

For second mortar, the substitution of cement by (5–10)% silica fume with the addition of super-plasticizer would enhance its properties tremendously. The underlying reason is that the silica fume is much finer than the cement and that led to the following:

i. Increase the density of mortar.
ii. Accelerate the rate of hardening.
iii. Increase workability.
iv. High compressive and tensile strength.
v. Strong bonding.
vi. High abrasion resistance.
vii. Controlled porosity and water permeability.
viii. Withstand ageing, weathering condition and freeze–thaw action.

9.3.2 Polymer-Modified Cement-Based Materials

Synthetic latexes are made by dispersing polymer particles in water to form a polymer emulsion. When the emulsion is added to Portland cement mortar or concrete, the spheres of polymer will come together to form a film that coats the aggregate particles and the hydrating cement grains and seals off voids, while still maintaining its permeability. The very important properties of polymer-modified cement-based mortars are that they have values for young modulus and coefficient of thermal expansion very similar to the base concrete.

Styrene–butadiene rubber (SBR) is a synthetic polymer used as an admixture for grout, mortar and concrete to enhance their performance as coating materials. For acrylic mortar, this generation of polymers has been developed by chemically combining the SBR and acrylic bases to form a range of polymeric admixtures for grout, mortar and concrete. The general properties of all these polymers-modified cement-based materials are described in the following:

i. Excellent bond strength.
ii. Superior adhesion in dry, damp and wet conditions to concrete.
iii. High flexural and tensile strengths.
iv. High abrasion resistance.
v. Low shrinkage.
vi. High resistance to chemicals, acids and alkalis.
vii. Enhanced resistance to freeze–thaw cycles.
viii. Low permeability to water and oils.

9.3.3 Epoxy-Based Materials

Moisture insensitive, modified and filler epoxy mortars and concretes would also be considered for testing in the laboratory. They are normally used at damaged surfaces where less than half an inch coating is required. The advantages of using epoxies are that they can produce and possess:

i. Excellent adhesion in dry, damp or wet conditions.
ii. High compressive and tensile strengths.
iii. Strong bonding to concrete.
iv. High abrasion and wear resistance.
v. High resistance to chemical attacks.

vi. On other hand, the use of fillers would improve the permeability, expansion, elastic properties and make them durable in relation to freeze–thaw cycles and on top of that would reduce the cost dramatically.

9.4 SELECTION OF REPAIR MATERIALS

The topic of repair is more complex than the design of new structures, and the management of rehabilitation is more complex than that of new construction [10]. The selection of an optimum repair material is one of the critical factors that dictate the success of any repair process. Surface preparation, the method of application, construction practices and inspection are the determining factors in the selection process. Selection of an optimum repair material with regard to cost, performance and risk is, however, not an easy task. It requires knowledge about the user expectations from the repair process, and the material behaviour in the cured and uncured states in the anticipated service and exposure conditions [11]. The entities that are involved in and affected by the repair process are the agencies that implement the repair process, the users of the facility and other users indirectly affected by the repair process. The agency's expectations from repair can be divided into two stages: (First stage: during the implementation of repair, second stage: after the repair is completed). During the implementation of repair, the agency's primary concern is the time required for completing the repair because this has a direct bearing on the user costs associated with the closure of the facility. Once the repair process is complete, the primary expectation of the agency is that the repair should be durable. This is indicated by the ability of the repaired pavement to endure varying environmental, temperature and load-related changes without deteriorating. Figure 9.1 shows a systematic approach that is required in the selection of a repair material, which accounts for all applicable parameters and their impacts on the choice between alternatives.

9.5 DEVELOPMENT OF GEOPOLYMER AS REPAIR MATERIALS

User-friendly geopolymer concrete can be used under conditions similar to those suitable for Ordinary Portland Cement concrete. These constituents of geopolymer paste and mortar shall be capable of being mixed with a relatively low alkali-activating solution and must be curable in a reasonable time under ambient conditions. The production of versatile, cost-effective geopolymer concrete can be mixed and hardened essentially like Portland cement [12]. Geopolymer concrete shall be used in repairs and rehabilitation works [13]. The previous work reveals that the geopolymer mechanical properties when compared with conventional repair materials gives different results and the compressive strength higher than conventional repair materials so the compressive strength of geopolymer mortar is enough for using as the repair mortar without the heat curing. The flexural strength and bond strength between geopolymer mortar and substrate are close to the commercial repair mortar at the age of 28 days. Hu et al. [14] found from the experiment done in the laboratory that the results reveal that the bond strength of geopolymeric repair materials have better repair characteristics than cement-based repair materials. However, the setting time of the geopolymer mortar is still longer than that of the commercial repair mortar. But

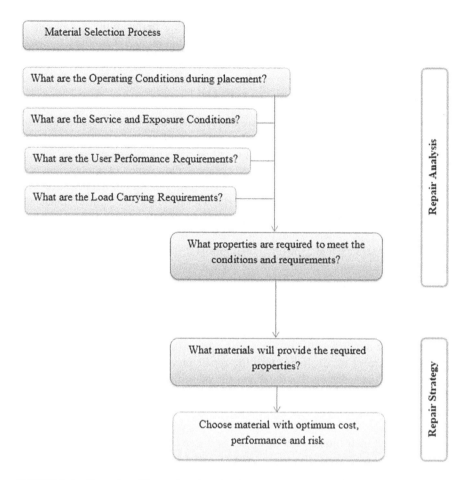

FIGURE 9.1 Flow chart illustrating the selection process for a repair material.

the results show the chloride penetration depth of the geopolymer mortar comparing with the commercial repair mortar and conventional cement mortar. As shown, the chloride penetration depth of all mixes of the geopolymer mortar is higher than that of the commercial repair mortar and conventional cement mortar at all ages. This is implied that the usage of the geopolymer mortar as the repair material still is not effective enough in terms of the durability. Therefore, to use geopolymer mortar as repair material, it must develop the chloride penetration resistance, bond strength and reduce the setting time.

The previous works provided the advantage of using geopolymer in concrete construction:

i. Cutting the world's carbon.
ii. The cost of fly ash is low.
iii. Better compressive strength.
iv. Fire proof.

Comparison of Geopolymer with Commercial Products

    v. Low permeability.
   vi. Eco-friendly.
  vii. Excellent properties within both acid and salt environments.
 viii. Greater corrosion resistance.
   ix. Substantially higher fire resistance (up to 2400°F).
    x. High compressive and tensile strengths.
   xi. Rapid strength gains and lower shrinkage.
  xii. Greenhouse gas reduction potential as much as 90% when compared with cement.

9.6 EFFICIENCY EVALUATION OF GEOPOLYMER AS REPAIR MATERIALS

Many methods and tests have been developed to evaluate the suitability of geopolymer binds as concrete surface repair materials. Physical and mechanical properties of durable tests were carried out such as bond strength with the existing concrete, resistance to abrasion–erosion, compressive strength, shrinkage–expansion before and after bonding, coefficient of thermal expansion, permeability and modulus of elasticity to evaluate the efficiency of geopolymer paste and mortars as concrete repair materials. In literature, many tests were considered to assess the bond strength in the interface between the geopolymers and existing concrete. Figure 9.2 schematically shows the test methods related to interface bond. While some of the methods are not so common in projects, some others such as "pull-off test" or "slant shear

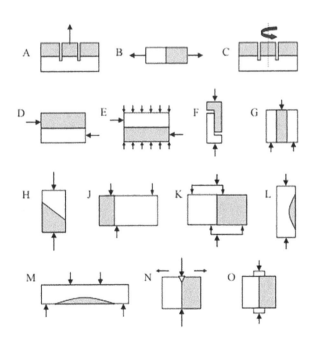

FIGURE 9.2 Various test methods to evaluate bond strength at the interface [15].

test" are more common and used extensively in projects [15]. Most of the standards and codes confirm these two tests. Less problems and shortcomings, being easy to set up and perform, wide range of applications and the reliability of the results are the main reasons why these two tests are more accepted.

Good bond strength between overlay and substrate is a key factor in the performance of concrete repairs. After repairing the concrete structure and replacing a new layer, there should be enough strength in both layers since the damaged part of substrate has been already removed and the new layer has been designed and placed according to the requirements of the work. Despite having adequate strength in both layers, the interface is still vulnerable to damages and could be the most sensitive part of the system.

Two layers have different modulus of elasticity, so exposed to the same load each shows different strains. The interface should be able to bear this difference. The same problem exists for the temperature strains. In addition, the new layer has shrinkage which is considered as another factor for interface weakness. Since the interface is the plane of discontinuity in the system, it exposes to all these extra forces and it should have enough resistance to hold the integrity of the layers. Thus, a key requirement of a repair material is to good adhesion at the interface. There are many factors which have influence on the bond strength and some test methods to figure out the strength and quality of the bond.

Most of these tests will be performed in accordance with ASTM standards. In a few cases, the standard specifications will be modified slightly to suit the local environmental conditions. In other cases, where no standard exists, other local or international standards will be used. In addition, the bond strength and abrasion–erosion resistance test will be conducted as the two pre-selection tests and will be used as screened tests to eliminate unsuitable mortars. The purpose of this screening is to reduce total number of tests, the total testing time and the cost. Figure 9.3 shows the procedure which was adopted to select the geopolymer mortars (GPMs) as concrete repair materials.

9.7 SUMMARY

Based on the above discussion, the following conclusions were drawn:

i. Bond strength, abrasion resistance and compressive strength are the most tests adopted to select the concrete surface repair materials.
ii. Compared to commercial repair materials, geopolymer paste and mortar presented excellent performance in terms of early strength, drying shrinkage resistance to elevated temperatures and durability performance.
iii. As there many factors effect on geopolymers performance as repair materials such as binder type, chemical composition of binders, alkaline solution properties and filler content, increasing the possibility to produce different types of geopolymer which suitable for different environments.
iv. Geopolymers containing high volume of ground blast furnace slag (GBFS) presented the optimum value of bond strength where an increase in the GBFS content could dissolve more silicate and led to the improvement of the reaction process to form C-S-H gel.

Comparison of Geopolymer with Commercial Products

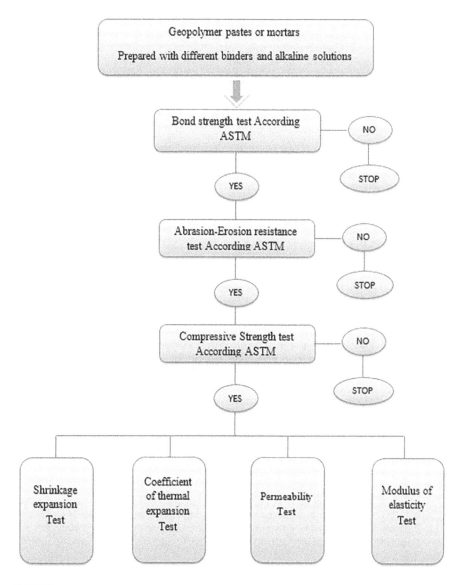

FIGURE 9.3 Flow chart of tests adopted to select GPMs as repair materials.

v. Increasing the content of fly ash (FA), palm oil fuel ash (POFA) and ceramic waste powder (CWP) up to 70% in ternary blended geopolymer matrix led reduced the bond strength more than 40% as compared to others levels.

vi. Most of geopolymer mixtures prepared with high volume of FA presented an excellent bond strength which was attributed to the low ratio of SiO_2 to Al_2O as compared to POFA and CWP matrixes.

vii. The bond strength of GPMs to normal concrete substrate in critical condition (30° slant shear) presented excellent results.

REFERENCES

1. Mirza, J., et al., Preferred test methods to select suitable surface repair materials in severe climates. *Construction and Building Materials*, 2014. **50**: pp. 692–698.
2. Garbacz, A., M. Górka, and L. Courard, Effect of concrete surface treatment on adhesion in repair systems. *Magazine of Concrete Research*, 2005. **57**(1): pp. 49–60.
3. Smoak, W.G., Polymer impregnation and polymer concrete repairs at Grand Coulee dam. *Special Publication*, 1985. **89**: pp. 43–50.
4. Gagné, R., et al. Repair of a reinforced concrete bridge deck using bond thin overlay: results of test shelf of bridge cosmos. in *ACI Annual Seminar*. 2003 Florida, pp. 26–27.
5. Gardner, N., Design provisions for shrinkage and creep of concrete design provisions for shrinkage and creep of concrete design provisions for shrinkage and creep of concrete. *Special Publication*, 2000. **194**: pp. 101–134.
6. Huseien, G.F. and K.W. Shah, Performance evaluation of alkali-activated mortars containing industrial wastes as surface repair materials. *Journal of Building Engineering*, 2020. **30**: p. 101234.
7. Huseien, G.F., K.W. Shah, and A.R.M. Sam, Sustainability of nanomaterials based self-healing concrete: An all-inclusive insight. *Journal of Building Engineering*, 2019. **23**: pp. 155–171.
8. Gagné, R., et al., Innovative concrete overlays for bridge-deck rehabilitation in Montréal, in *Concrete repair, rehabilitation and retrofitting II*. Edited by Mark G. Alexander, Hans-Dieter Beushausen, Frank Dehn, Pilate Moyo, 2008: CRC Press. pp. 369–370.
9. Huseien, G.F., et al., Geopolymer mortars as sustainable repair material: a comprehensive review. *Renewable and Sustainable Energy Reviews*, 2017. **80**: pp. 54–74.
10. Sykora, M. and M. Holicky, Reliability assessment of industrial heritage buildings. *Engineering Mechanics*, 2012. **18**: p. 100.
11. Rajabipour, F., et al., *Longitudinal cracking in concrete at bridge deck dams on structural rehabilitation projects*. Department of Transportation, Philadelphia, USA. 2012: pp. 1–232.
12. Zou, F., et al., Enhancement of early-age strength of the high content fly ash blended cement paste by sodium sulfate and C–S–H seeds towards a greener binder. *Journal of Cleaner Production*, 2020. **244**: p. 118566.
13. Aleem, M.A. and P. Arumairaj, Geopolymer concrete–a review. *International Journal of Engineering Sciences & Emerging Technologies*, 2012. **1**(2): pp. 118–122.
14. Hu, S., et al., Bonding and abrasion resistance of geopolymeric repair material made with steel slag. *Cement and Concrete Composites*, 2008. **30**(3): pp. 239–244.
15. Silfwerbrand, J., Shear bond strength in repaired concrete structures. *Materials and Structures*, 2003. **36**(6): pp. 419–424.

10 Sustainability of Geopolymer as Repair Materials

10.1 INTRODUCTION

The Ordinary Portland Cement (OPC) industry is responsible for 5%–7% of all CO_2 emissions generated by human activities [1]. Continuing cement production at the current rate may cause irreparable damage to global ecological systems. Thus, the development of eco-efficient alternatives to OPC is of utmost importance. Moreover, efficient industrial waste management and reducing the consumption of non-renewable natural resources are vital for sustainable development and cleaner environment. Since the disposal of industrial waste materials is often associated with adverse environmental impacts, a wide range of the so-called "green" concrete and mortar mixtures incorporating industrial by-products has been developed [2–7]. Since OPC is the primary concrete constituent responsible for CO_2 emissions and embodied energy (EE), effort has been made to fully or partially replace it by supplementary cementitious materials (SCMs). Industrial by-products and agriculture wastes such as fly ash (FA) also known as pulverized fuel ash in the United Kingdom that is acquired from power station after firing the coal, ground-granulated blast furnace slag (GBFS) which is obtained by quenching molten iron slag (a by-product of iron and steel-making) from a blast furnace in water or steam, waste ceramic powder (WCP) which is released as wastes from ceramic and construction industries, palm oil fuel ash (POFA) that is produced from the palm oil fibres, bunches and shells as fuel for power generation in the mills, rice husk ash and sugarcane bagasse ash, have been considered as SCMs for full or partial replacement of OPC.

Geopolymer paste, mortar and concrete which are manufactured using industrial by-products have demonstrated eco-efficient features, while achieving appropriate mechanical strength and durability. Generally, such mortars and concretes are prepared using starting source materials rich in silicon (Si), aluminium (Al) and calcium (Ca) with alkali activation (such as sodium silicate or/and sodium hydroxide). The compatible nature of aluminium-substituted calcium–silicate–hydrate (C-(A)-S-H) and sodium–aluminium–silicate (N-A-S-H) gels has significant influence on the geopolymer mortars (GPMs) and alkaline solution-activated alumina-silicate systems, wherein both products may be obtained compared to calcium–silicate–hydrate (C-S-H) gel with OPC. They allow full replacement of OPC primarily using SCMs in their formulation, thus resulting in OPC-free concrete and mortar. Literature study on GPMs showed excellent properties such as high early strength, more resistance to aggressive environments and lower pollution compared to cement mortar. The most commonly used materials in GPM manufacturing are FA, metakaolin (MK) and GBFS as reported in [8–10]. Although previous research has confirmed the excellent properties of FA and GBFS–based GPMs, the very nature of these industrial by-products implies varying mineralogical and chemical composition, making the standardization process to reach desirable mechanical and durability properties difficult. Moreover, GPMs do not require the clinker manufacturing process needed for OPC at 1350°C–1450°C, but are rather produced at relatively low temperatures of 25°C–100°C. This leads to a substantial reduction in CO_2 emissions resulting from decarbonization of limestone and in the EE needed in clinker production. Nevertheless, to ensure reliable mechanical properties and environmental benefits of GPMs, appropriate life cycle assessment (LCA) is undeniable. LCA is a reliable standardized methodology to evaluate the environmental features of GPMs and demonstrate rationally the representation of an effective and viable alternative to OPC. Considering that the pertinent results reported in the open literature remain contradictory, the environmental impact of GPMs remains controversial and open to debate [11–14].

LCA is a lucid method that evaluates the environmental impact of products over their life cycle, providing precise and scientifically based results [15–17]. It is therefore a rational and robust tool for assessing the ecological feasibility of incorporating recycled wastes and industrial by-products into green concrete production [18]. To investigate the environmental impact of concrete manufacturing, it is necessary to evaluate the entire life cycle, from the extraction of raw materials to the final waste disposal stage. Attention should be paid to the environmental features of OPC substitution with alternative industrial by-products in view of sustainability considerations. Previous literature has primarily focused on the mechanical properties and durability of alternative binder materials in concrete, but did not generally consider comprehensive LCA to environmentally justify cement substitution as reported in [19–22]. Therefore, adopting the LCA method for the replacement of OPC concrete with eco-efficient alternatives is essential [14]. Specific parameters in green concrete and mortar design should consider obtaining adequate workability, mechanical strength, durability, cost, and aesthetics, along with enhanced environmental footprint.

Accordingly, the present study investigates the compressive strength (CS), durability and the environmental impact of alkali-activated mortars composed of industrial by-products. CO_2 emissions and EE which represent fundamental parameters in the cradle-to-gate LCA were investigated in detail for 42 ternary blended GPM mixtures. Using the available experimental test database, an optimized Artificial Neural Network (ANN) combined with the cuckoo optimization algorithm (COA) was developed to estimate the CO_2 emissions and EE of GPMs. This research contributes significantly towards the implementation and standardization of industrial scale manufacturing approaches of low carbon footprint GPM mortars in the foreseeable future, particularly in geographic locations with presence of volcanic ashes, and East Asian countries with extensive production of FA and POFA. Furthermore, the final weights and biases of the trained ANN can be used to design GPMs with targeted mechanical properties and CO_2 emissions based on locally available industrial by-products.

10.2 GEOPOLYMER PREPARATION

In this experiment, the pure GBFS obtained from Ipoh, Malaysia by a supplier and used as received without any further treatment is utilized as the main resource calcium materials in Alkali-activated mortars (AAMs) production. From power station, Johor, Malaysia, the low-calcium FA was collected and used as received. Raw POFA is collected from the local palm oil industry (Malaysia). Incomplete combusted fibres and kernel shells are separated using a 300 μm sieve before being dried in an oven for at least 24 hours at 105°C±5°C to remove moisture. The POFA is grounded using Los Angeles machine to obtain a particle size of 10 μm. To achieve the desired level of fineness, the POFA is crushed for 12,600 cycles over 6 hours. From White Horse ceramic manufacturer in Johor, Malaysia, homogeneous waste tile ceramics were collected which is same in thickness with no glassy coating. They were crushed using jaw crusher and after that they were sieved with 600 μm to remove big size particles. The ceramic waste particles that passed through 600 μm sieve were ground using Los Angeles abrasion machine with 20 stainless steel balls of 40 mm in diameter for 6 hours and finally known as waste ceramic powder (WCP). FA, POFA and WCP are used as a resource of aluminosilicate material for making GPMs. Colours of GBFS, FA, POFA and WCP were off-white, light grey, dark grey and light grey, respectively. From their physical properties, the lower specific gravity was observed with POFA (1.96) compared to 2.2, 2.6 and 2.9 of FA, WCP and GBFS, respectively. The medium particle size of GBFS, FA, POFA and WCP were 12.8, 10, 8.2 and 35 μm, respectively.

Using X-ray fluorescence spectroscopy (XRF, HORIBA, Singapore, Singapore), the chemical compositions of the industrial by-product materials were determined (Table 10.1). It was revealed that the main compound in POFA, FA, and WCP was SiO_2 (64.2%, 57.2%, and 72.6%, respectively), whereas in GBFS, CaO was the main compound (51.8%). Al_2O_3, SiO_2 and CaO are the essential oxides throughout the hydration and production processes of the C-(A)-S-H gels. Nevertheless, the low contents of Al_2O_3 and CaO in WCP require adding materials comprising high quantities

TABLE 10.1
Physical and Chemical Composition of Industrial By-product Materials Used

Material	GBFS	FA	POFA	WCP
Specific gravity	2.9	2.2	1.96	2.6
Avr. particle size (μm)	12.8	10	8.2	35
SiO_2	30.8	57.20	64.20	72.6
Al_2O_3	10.9	28.81	4.25	12.6
Fe_2O_3	0.64	3.67	3.13	0.56
CaO	51.8	5.16	10.20	0.02
MgO	4.57	1.48	5.90	0.99
K_2O	0.36	0.94	8.64	0.03
Na_2O	0.45	0.08	0.10	13.5
SO_3	0.06	0.10	0.09	0.01
LOI	0.22	0.12	1.73	0.13

of Al_2O_3, such as FA, and CaO-rich materials, such as GBFS, to produce high-performance alkali-activated binders. According to ASTM C618-15 [23], FA and WCP are classified as Class F pozzolans due to the existence (higher than 70%) of $SiO_2 + Al_2O_3 + Fe_2O_3$.

Ternary blended GPMs were examined to determine the influence of calcium oxide on the geopolymerization process. Using trial mixes, the optimum ratio of sodium silicate-to-sodium hydroxide alkali activators, sodium hydroxide molarity, binder-to-aggregate ratio and alkaline solution-to-binder ratio were determined as 0.75, 4 M, 1, and 0.4, respectively, where these values were fixed for all GPMs. Analytical-grade sodium silicate solution "Na_2SiO_3" (NS), comprising SiO_2 (29.5 wt %), Na_2O (14.70 wt %) and H_2O (55.80 wt %(in combination with sodium hydroxide (NaOH), was used as the alkali activator to prepare the proposed GPM mixtures. The NaOH pellet was dissolved in water to make the alkaline solution with 4 M concentration. In the first phase, the solution was cooled for 24 hours and then added to the sodium silicate (NS) solution to obtain an alkaline activator solution with a modulus ratio ($SiO_2:Na_2O$) of 1.02. The ratio of NS-to-NaOH was fixed to 0.75 for all the alkaline mixtures. Four ternary blended AAMs were investigated, where at each level, the GBFS percentage, as a source of CaO, remained constant at a minimum of 20% in the replacement process and a maximum of 70%, as given in Table 10.2.

TABLE 10.2
Ternary Blended GPM Designs and Calculated EE and CO_2 Emissions

GPM Designs	Binder Constitution (Composed of Industrial Waste Materials)				Sustainable and Mechanical Features		
	FA	GBFS	WCP	POFA	EE (MJ/m³)	CO_2 Emission (kgCO_2/m³)	28-Days CS (MPa)
				High-volume FA mix design			
1	0.70	0.30	0.00	0.00	709.00	39.55	78.18
2	0.70	0.20	0.00	0.10	699.00	36.68	65.89
3	0.60	0.40	0.00	0.00	859.00	47.05	80.51
4	0.60	0.30	0.00	0.10	849.00	44.18	81.70
5	0.60	0.20	0.00	0.20	839.00	41.30	52.60
6	0.50	0.50	0.00	0.00	1009.00	54.55	80.46
7	0.50	0.40	0.00	0.10	999.00	51.68	76.90
8	0.50	0.30	0.00	0.20	989.00	48.80	70.40
9	0.50	0.20	0.00	0.30	979.00	45.93	46.24
				High-volume POFA mix design			
10	0.00	0.30	0.00	0.70	1689.00	71.93	34.53
11	0.10	0.20	0.00	0.70	1539.00	64.43	23.04
12	0.00	0.40	0.00	0.60	1699.00	74.80	45.96
13	0.10	0.30	0.00	0.60	1549.00	67.30	37.80
14	0.20	0.20	0.00	0.60	1399.00	59.80	28.80

(Continued)

TABLE 10.2 (Continued)
Ternary Blended GPM Designs and Calculated EE and CO₂ Emissions

GPM Designs	Binder Constitution (Composed of Industrial Waste Materials)				Sustainable and Mechanical Features		
	FA	GBFS	WCP	POFA	EE (MJ/m³)	CO_2 Emission (kgCO_2/m³)	28-Days CS (MPa)
15	0.00	0.50	0.00	0.50	1709.00	77.68	55.64
16	0.10	0.40	0.00	0.50	1559.00	70.18	47.10
17	0.20	0.30	0.00	0.50	1409.00	62.68	40.60
18	0.30	0.20	0.00	0.50	1259.00	55.18	36.80
			High-volume GBFS mix design				
19	0.30	0.70	0.00	0.00	1309.00	69.55	85.09
20	0.20	0.70	0.00	0.10	1449.00	74.18	97.75
21	0.10	0.70	0.00	0.20	1589.00	78.80	86.40
22	0.00	0.70	0.00	0.30	1729.00	83.43	70.53
23	0.40	0.60	0.00	0.00	1159.00	62.05	80.68
24	0.30	0.60	0.00	0.10	1299.00	66.68	72.44
25	0.20	0.60	0.00	0.20	1439.00	71.30	71.93
26	0.10	0.60	0.00	0.30	1579.00	75.93	70.84
27	0.00	0.60	0.00	0.40	1719.00	80.55	70.22
28	0.50	0.50	0.00	0.00	1009.00	54.55	80.46
29	0.40	0.50	0.00	0.10	1149.00	59.18	80.43
30	0.30	0.50	0.00	0.20	1289.00	63.80	67.22
31	0.20	0.50	0.00	0.30	1429.00	68.43	65.14
32	0.10	0.50	0.00	0.40	1569.00	73.05	56.34
33	0.00	0.50	0.00	0.50	1709.00	77.68	55.64
			High-volume WCP mix design				
34	0.00	0.30	0.70	0.00	1323.81	58.66	34.02
35	0.10	0.20	0.70	0.00	1173.81	51.16	22.40
36	0.00	0.40	0.60	0.00	1385.98	63.43	68.44
37	0.10	0.30	0.60	0.00	1235.98	55.93	52.08
38	0.20	0.20	0.60	0.00	1085.98	48.43	46.76
39	0.00	0.50	0.50	0.00	1448.15	68.20	74.12
40	0.10	0.40	0.50	0.00	1298.15	60.70	66.19
41	0.20	0.30	0.50	0.00	1148.15	53.20	60.17
42	0.30	0.20	0.50	0.00	998.15	45.70	56.47
Average					1292.03	61.39	61.30
STDEV					293.71	12.26	18.70

10.3 STRENGTH PERFORMANCE

At 365 days of curing age (laboratory temperature of 27°C ± 1.5°C and relative humidity of 75%), the compression strength test was carried out, following ASTM C109-109M. Three samples were tested for this age; after preparation, each sample was placed precisely between the top and bottom metal-bearing plates, in line with the relevant standard specifications. A consistent loading rate of 2.5 kN/s was applied to the samples. Density and CS figures, based on the weight and size of the samples, were automatically created due to the machine's construction. The average values of three readings were adopted for each batch.

Table 10.2 shows the CS for all 42 GPMs mixtures. It can be observed that the highest mechanical properties were achieved by GPMs made with a high volume of GBFS, while GPMs made with a high volume of POFA resulted in the lowest mechanical properties. The mechanical properties in GPMs made with a high volume of WCP also were not satisfactory. However, increasing the GBFS dosage in the binder mass improved the mechanical strength in this category. The mechanical features in GPMs made with a high-volume FA were significantly dependent on the percentage of GBFS in the binder mass, where substituting GBFS by POFA significantly decreased the CS. Overall, the average CS of the studied GPMs mixtures was 61.3 MPa, which is satisfactory, while having much lower EE and CO_2 emission compared to traditional OPC-based mortars.

Generally, sulphuric acid attacks GPMs by dissolving the binder paste matrix, leading to the weakening of mechanical properties of the GPM mortar. In this research, using deionized water, a 10% H_2SO_4 acid solution was prepared, and its effects on the residual CS, mass loss and ultrasonic pulse velocity (UPV) of GPMs were investigated at the age of 28 and 365 days in compliance with ASTM C267 specifications [24]. To sustain the pH of the solution over the span of the test duration, it was changed every 2 months. Sulphate attack on the studied alkali-activated mortar specimens was caused by the sulphate ions $(SO_4)^{2-}$ that were transmitted into the mortar from varying concentrations in the water together with magnesium, calcium or sodium cations. Magnesium sulphate solution was also employed to evaluate the resistance to sulphate attack of the alkali-activated specimens using a test procedure similar to that adopted for the sulphuric acid attack test.

Figure 10.1 illustrates the residual CS and mass loss of all the 42 GPM mixtures after 365 days of immersion in the sulphuric acid solution. On average, the CS and specimen mass declined by 90%–0.56%, respectively, compared to the control intact specimens. The maximum reduction in CS was inflicted to specimens in the category of high-volume GBFS by around 300%, while the specimens with high-volume WCP experienced major mass loss of an average 0.85%. Figure 10.2 illustrates the residual CS and mass loss of all 42 GPMs after 365 days of immersion in the sulphate solution. There was generally a similar pattern observed for residual CS compared to that of immersion in the sulphuric acid solution, whereas the maximum mass loss was recorded for specimens with high-volume GBFS by an average of 0.66%.

Figure 10.3 displays the physical appearance of the cubic GPM mortar specimens prepared with different industrial by-products after 365 days of immersion in the sulphuric acid and sulphate solutions. Comparing Figure 10.3b (after 365 days of immersion in sulphuric acid solution) to the control intact samples (Figure 10.3a), it can be observed that the durability of GPM specimens exposed to the sulphuric acid environment gradually decreased with increasing GBFS content. However, increasing the level of FA, POFA and WCP from 30% to 70% led to increased

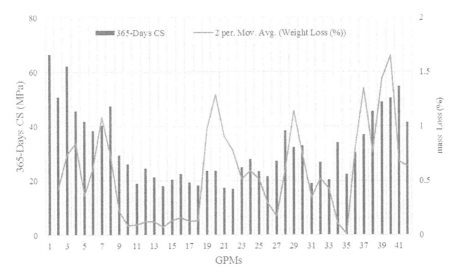

FIGURE 10.1 Effects of exposure to sulphuric acid solution on CS and mass of GPMs.

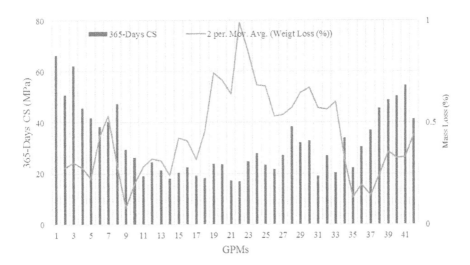

FIGURE 10.2 Effects of exposure to sulphate solution on CS and weight of GPMs.

resistance of GPM specimens to sulphuric acid attack, showing an excellent durability performance. Upon exposure of the GPM specimens to the sulphuric acid solution, the $Ca(OH)_2$ compound in the mortar reacted with SO_4^{-2} ions and formed gypsum ($CaSO_4.2H_2O$). This caused expansion in the alkali-activated matrix and additional cracking in the interior of the specimens, as indicated through visual appearance of these specimens. The high calcium oxide in the high-volume GBFS

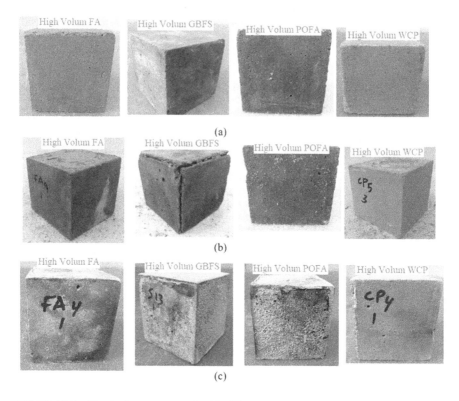

FIGURE 10.3 Physical appearance of cubic GPM specimens made with different industrial by-products. (a) control sample. (b) After 365 days of immersion in sulphuric acid solution. (c) After 365 days of immersion in sulphate solution.

geopolymer specimens compared to that in the other matrixes resulted in more abundant gypsum formation.

Therefore, degradation in residual CS along with more substantial mass loss was observed for GPM specimens made with high-volume GBFS and immersed in the sulphuric acid solution. A reduction in mass loss can be explained by increasing the SiO_2 and Al_2O_3 contents associated with a reduction of the CaO content. Moreover, decreasing the CaO content reduced gypsum formation, thus increasing the durability of the alkali-activated mortar specimens (Figure 10.3).

For the durability to the sulphate environment, it was found that increasing the FA, POFA and WCP levels in the geopolymer matrix mitigated the deterioration (Figure 10.3c) and increased the residual strength. Several researchers have reported sulphate deterioration can cause mechanical strength loss, expansion, spalling of surface layers and ultimately disintegration. Most experts attribute sulphate attack to the formation of expansive ettringite ($3CaO \cdot Al_2O_3 \cdot 3CaSO_4 \cdot 32H_2O$) and gypsum ($CaSO_4 \cdot 2H_2O$), which may be accompanied by expansion or softening.

UPV testing can be deployed in situ as a non-destructive evaluation technique to check the quality of concrete in terms of material discontinuities, and damages such

Sustainability of Geopolymer

as cracks and delamination under a given exposure time. In this test, the strength and quality of concrete are appraised by measuring the velocity of an ultrasonic pulse passing through the concrete element. The pulse velocity can be determined by measuring the length between the transducers and the travel time, as per Eq. 10.1 where, x is distance and t is transit time, where more rapid velocity indicates better material integrity, higher density and superior quality of the material.

$$\text{UPV} = v_c(x, t) = x/t \tag{10.1}$$

The experimental results confirm that pulse velocity decreased by an average of about 8% and 5% for specimens immersed into the acid and sulphate solutions for a period of 365 days, respectively. In this study using nonlinear regression analysis, an exponential function for estimating the relationship between CS and pulse velocity of GPMs was established. Figure 10.4 depicts the relationship between the mean values of UPV and CS for all the 42 GPM mixtures investigated before and after 365 days of immersion in sulphuric acid and sulphate solutions. The results confirm that there was an inverse correlation between CS and pulse velocity reduction, where GPMs with lower CS have shown larger reduction of pulse velocity. The highest pulse velocity before and after immersion in the sulphuric acid and sulphate solutions was achieved by GPM mixture 20 made with 20% FA + 70% GBFS + 10% POFA, with a CS of 97.75 MPa. Generally, GPMs with higher dosage of GBFS exhibited the highest value of pulse velocity before and after immersion in the sulphuric acid sulphate solution compared to other mixtures. However, GPMs incorporating high-volume WCP demonstrated appropriate performance in resisting exposure to the sulphate solution, where the average pulse velocity and CS remained nearly unchanged before and after the exposure. This can be explained by the morphology of this alkali-activated mixture which possess high magnesium sulphate ($MgSO_4$) and silicon dioxide SiO_2 contents, providing resistance against sulphate attack. Previous literature indicated the relationship between CS and pulse velocity as a measure of material deterioration, internal cracking and pre-existing defects in mortars before and after immersion in sulphuric acid and sulphate solutions using the following exponential function [25–27], where V is the UPV, and the coefficients A and B are empirical constants.

$$CS = Ae^{(BV)} \tag{10.2}$$

10.4 LIFE CYCLE ASSESSMENT

In this research, the main objective of LCA was to contrast the production of GPMs manufactured with ternary blended industrial by-products with a benchmark conventional OPC-based mortar. The LCA was concerned with CO_2 emissions and EE in compliance with the Inventory of Carbon and Energy (ICE) [28] in which the system boundary begins with the raw material acquisition (cradle) and ends at the factory gate, exclusive of the impacts associated with transportation, service or use life, and end-of-life. Table 10.3 shows the CO_2 emissions and EE for all binder materials, OPC and fine aggregate as provided by ICE.

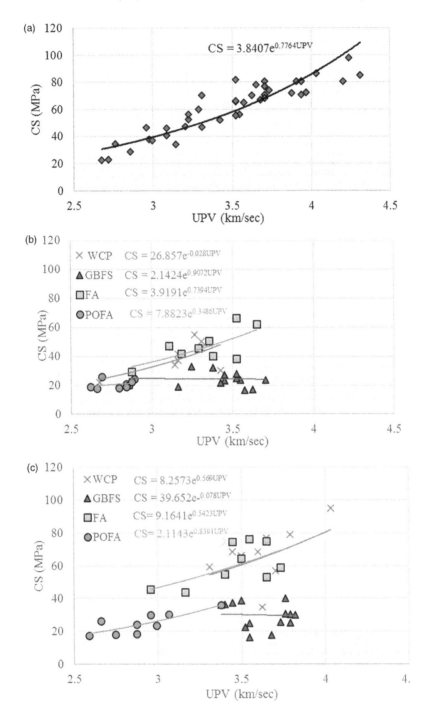

FIGURE 10.4 Relationship between UPV and CS for all 42 AAM mixtures. (a) Original condition. (b) After immersion in sulphuric acid solution. (c) After immersion in sulphate solution.

TABLE 10.3
Assumptions Used in LCA Calculation (Data Retrieved from) [28]

Material	CO_2 Emission (kg CO_2/kg)	EE (MJ/kg)
POFA	0.0542	1.5
FA	0.008	0.10
GBFS	0.083	1.6
WCP	0.0353	0.9783
Fine Aggregate	0.0048	0.081
OPC	0.73	4.50

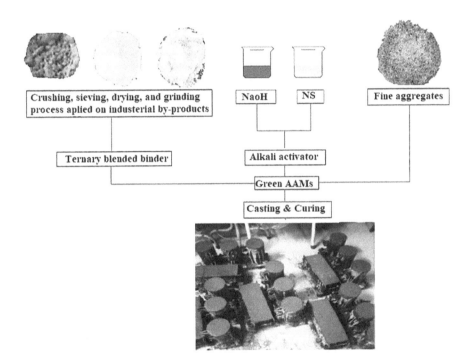

FIGURE 10.5 Cradle-to-gate and production stage of GPMs.

Figure 10.5 depicts the cradle-to-gate and production stage of ternary blended AAMs. Such conventional cradle-to-gate method was already widely applied to "green" concrete containing industrial by-products, AMMs and geopolymers made with FA, GBFS, MK, along with alkali activators such as sodium hydroxide [29–31]. Nevertheless, such a traditional method is not sufficiently reliable for assessing the environmental impacts (or benefits) of green concrete products since it precludes the advantageous effects of alkali-activated binders composed of industrial by-products on the mechanical properties and durability. Only a dearth of research considered the normalization of the climate change potential with respect to green concrete's mechanical properties [32]. Since the life span of concrete and mortar can be extended by improving the durability and mechanical properties, these parameters should be captured in the domain of the LCA criteria. Thus, in the present research, the cradle-to-gate LCA is adapted by taking the mechanical properties and durability of the GPMs into consideration. Using this approach, not only the impact of material manufacturing is accounted for but also the impacts of service life impacts are incorporated in the LCA criteria.

The cradle-to-gate LCA considered in the current research is characterized by the major processes associated with raw material extraction and material production stages in compliance with ICE. The functional unit of CO_2 emissions and EE is per cubic metre of GPM. Additionally, a revised cradle-to-gate system boundary was applied to the GPMs to include the service life phase on the basis of performance criteria. Accordingly, the service life impacts were incorporated through consideration of the CS and durability (sulphuric acid and sulphate resistance) of GPMs. The following equation was considered to estimate the CO_2 emission and EE per cubic metre of the GPMs:

$$Total\ CO_2\ emission\ or\ EE = \sum_{i=1}^{n} mi(pi) \qquad (10.3)$$

where the left-hand side of the equation indicates the net amount of CO_2 emission (kg CO_2) and EE (MJ) for every cubic metre of GPM production, mi indicates the fraction of component i and pi specifies the CO_2 emissions (kg) and EE (MJ) of per cubic metre of component i produced.

The estimated CO_2 emissions and EE per cubic metre of GPM for all the 42 mixtures explored are illustrated in Figures 10.6 and 10.7, respectively. The percentage distribution of CO_2 emissions and EE associated with the production of non-cementitious materials, fine aggregate, mixing and alkali activator was considered constant for all the GPM mixtures. The results indicate that the GPM mixture with high-volume FA emitted the least amount of CO_2 and consumed the least amount of energy with an average of 45.5 kg CO_2/m^3 and 881.2 MJ/m^3, respectively. On the other hand, the GPM mixture made with high-volume GBFS emitted the highest CO_2 amount, while the GPM mixture made with high-volume POFA consumed the highest amount of energy with an average of 70.6 kg CO_2/m^3 and 1534.5 MJ/m^3, respectively. The results confirm that the CO_2 emissions and energy consumption associated with the production of GPM made with GBFS and POFA are relatively higher compared to that GPM made with other industrial waste materials. Such results can be explained by the higher amount of electricity required for grinding GBFS to obtain the recommended particle size and for drying POFA in the oven at a temperature of 110°C ± 5°C for 24 hours. Overall, it can be concluded that the highest CO_2 emissions and EE of all 42 GPM mixtures studied were significantly lower than that of the benchmark conventional mortar prepared using OPC (1/3 cement–sand mix) which is associated with 436.8 kg CO_2/m^3 and 2793 MJ/m^3, respectively.

Sustainability of Geopolymer

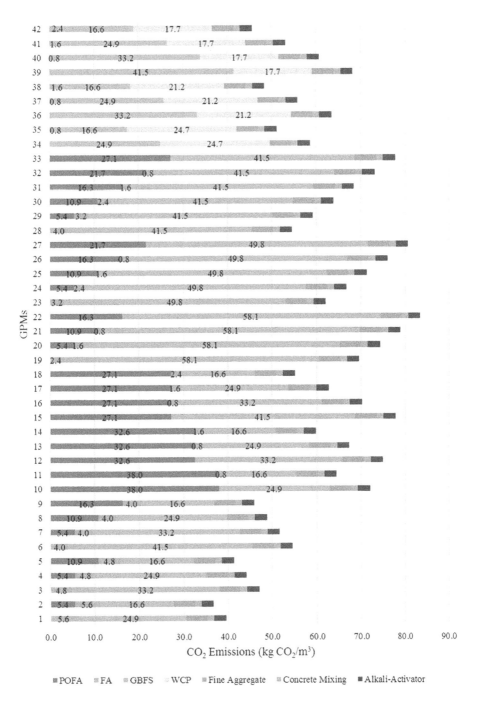

FIGURE 10.6 Distribution of CO_2 emission by GPM ingredient and phase.

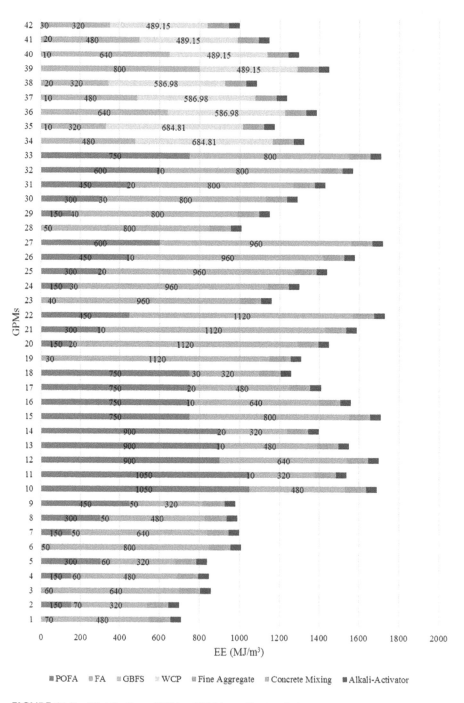

FIGURE 10.7 Distribution of EE by GPM ingredient and phase.

10.5 MODIFIED LCA WITH RESPECT TO CS AND DURABILITY

For a consistent and systematic comparison among GPM mixtures, their CO_2 emissions are normalized with respect to CS (Figure 10.8). The results confirm that in GPM mixtures incorporating high-volume FA and GBFS, lower intensity of normalized CO_2 emissions is achieved. The highest intensity of normalized CO_2 emissions was recorded for GPM mixtures containing high-volume POFA, which is correlated to its relatively low CS and high CO_2 emission. For a given CS, lower intensity of normalized CO_2 emissions can be achieved by replacing FA with GBFS. For instance, at CS of around 80 MPa, a reduction in GBFS from 70% (Mixture 9) to 30% (Mixture 1) reduced the normalized CO_2 emission from 1.15 to 0.5 $CO_2.m^{-3}$/MPa. By substituting 20% of WCP mass (Mixture 35) with FA (Mixture 41) in GPMs containing high-volume WCP, the intensity of normalized CO_2 emissions could be decreased by around two times.

To include the durability in the performance criteria of the studied GPM mixtures, their CO_2 emissions were normalized with respect to CS after 365 days of immersion in the sulphuric acid and sulphate solutions (Figures 10.9 and 10.10). The results confirm that the normalized CO_2 emissions for GPM mixtures made with high-volume POFA and GBFS were relatively higher than that for other mixture designs. This can ascribed to the fact that the mixtures containing GBFS and POFA were vulnerable to sulphuric acid and sulphate attack, where their CS significantly decreased after 365 days of immersion in these solutions. The highest normalized CO_2 emission in

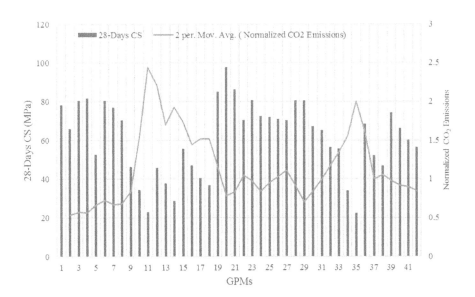

FIGURE 10.8 Illustration of CS versus normalized CO_2 emissions.

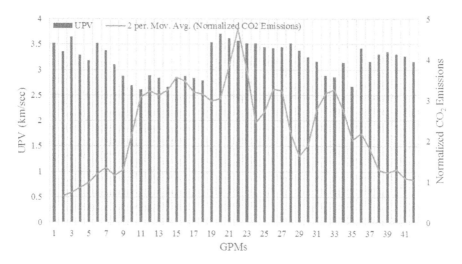

FIGURE 10.9 UPV versus normalized CO_2 emission subjected to sulphuric acid attack.

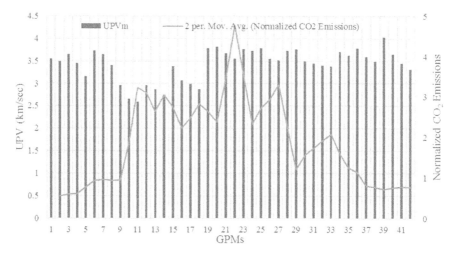

FIGURE 10.10 UPV versus normalized CO_2 emission subjected to sulphate attack add unit for UPV.

both sulphuric acid and sulphate attack was achieved by the GPM mixture 22 incorporating high GBFS and POFA, with an intensity of around 5 $CO_2.m^{-3}$/MPa, which is nearly five times higher compared to that of the intact control condition. The intensity of normalized CO_2 emissions in GPMs made with a high percentage of GBFS and POFA did not experience major changes compared to their original intact conditions, where the average intensity in the sulphuric acid and sulphate attack exposures were around 1.26 and 0.83 $CO_2.m^{-3}$/MPa, respectively.

10.6 ANN FOR ESTIMATING CO_2 EMISSION AND EE

10.6.1 RATIONALE

An ANN combined with a metaheuristic algorithm was developed to estimate CO_2 emissions and EE of GPM mixtures. The model's final weight and bias values can be used to design GPM mixtures with targeted CO_2 emissions and energy consumption based on available local waste materials. The multilayer feed-forward network provides a reliable feature for ANN structures and was thus used in this research. This network comprises three individual layers: the input layer, where the data are defined to the model; the hidden layer/s, where the input data are processed and finally, the output layer, where the results of the feed-forward ANN are produced. Each layer contains a group of nodes referred to as neurons that are connected to the proceeding layer. The neurons in the hidden and output layers consist of three components: weights, biases, and an activation function that can be continuous, linear or nonlinear. Standard activation functions include nonlinear sigmoid functions (logsig, tansig) and linear functions (poslin, purelin) [33]. Once the architecture of a feed-forward ANN (number of layers, number of neurons in each layer, activation function for each layer) is selected, the weight and bias levels should be adjusted using training algorithms. One of the most reliable ANN training algorithms is the backpropagation (BP) algorithm, which distributes the network error to arrive at the best fit or minimum error [34,35] and was used accordingly in this study.

10.6.2 CUCKOO OPTIMIZATION ALGORITHM

Bird species lay eggs for reproduction. Finding a safe nest to lay and hatch their eggs and raise the chicks to the point of independence is always a challenge for birds. Therefore, birds have been using different approaches, including intricate design, artistry and complex engineering so that even the all-seeing eyes had hardly ever found them. Other birds give out with every conventional of parenthood and homemaking and rely on a gimmick to raise the young. These categories of birds, the so-called "brood parasites," lay their eggs in the nest of other species instead of building their own nests, leaving those parents to take care of their chicks. A well-known brood parasite is cuckoo, a skilful in the art of cruel deception [36]. The cuckoo starts with an initial population. They have some eggs that they will lay in the nest of several host birds. Its strategy involves speed, stealth and surprise, where the mother takes away one egg laid by the host, lays her own egg. They carefully imitate the pattern and colour of their own eggs to match that of their hosts. Some of these eggs, which are more similar to the host bird's eggs, will have a better chance of growing and becoming an adult cuckoo. Other eggs are detected and destroyed by the host bird. The number of eggs grown indicates the suitability of the nests in that area. The more eggs that can survive in an area, the more profit (desire) will be allocated. Therefore, the situation in which the largest number of eggs is saved will be a parameter that they intend to optimize [36].

10.6.3 GENERATION OF TRAINING AND TESTING DATA SETS

To train and develop a reliable ANN, the chemical properties of the industrial by-products, see Table 10.1, were taken into account on the basis of input variables. The input and output variables along with their properties are shown in Table 10.4. It can be observed in this Table that the number of input and output variables are 8 and 2, respectively. Since a large number of input parameters in ANN generally tends to increase the error, Principal Component Analysis (PCA) was considered to make the input parameters orthogonal to each other. Accordingly, the input density diagram is shown in Figure 10.11.

PCA is a dimension-reduction tool that can be used to reduce a large set of variables to a small set that still contains most of the information in the original large set.

TABLE 10.4
Characteristics of Studied Input and Output Parameters

Parameters	Type	Unit	Max	Min	STD	Average
FA	Input	Mass (%)	0.70	0.00	0.21	0.25
GBFS	Input	Mass (%)	0.70	0.20	0.16	0.41
CWP	Input	Mass (%)	0.70	0.00	0.24	0.12
POFA	Input	Mass (%)	0.70	0.00	0.23	0.22
$SiO_2:Al_2O_3$	Input	Ratio	8.63	2.10	1.58	4.03
$CaO:SiO_2$	Input	Ratio	0.97	0.17	0.23	0.52
$CaO:Al_2O_3$	Input	Ratio	4.41	0.66	1.09	2.05
Age	Input	Day	28.00	1.00	10.79	9.75
EE	Output	MJ/m^3	1729.00	699.00	291.06	1292.03
CO_2 emission	Output	$kgCO_2/m^3$	83.43	36.68	12.15	61.39

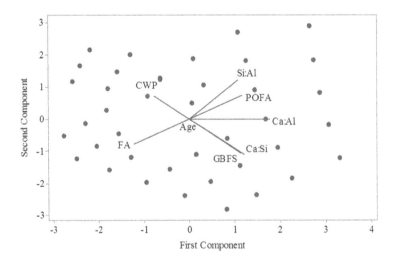

FIGURE 10.11 Scatter graph of the total density of input parameters using PCA.

This can be achieved by applying a transformation function, the so-called Principal Components (PC), on the primary variables. PCs are unrelated to each other and are sorted in such a way that the primary variables contain the most features of variances of the primary variables. The detailed information of this method can be found in [37,38]. Using PCA, Table 10.5 shows the influence of each parameter on inputs variables. It can be observed in this Table that the conversion of eight input parameters into four variables, PCA 1 to PCA 4, resulted in using 98.8% of the data, and as a consequence of such data convergence, better model results can be obtained. The resulting input variables using PCA are shown in Table 10.6.

Therefore, according to the optimal accuracy of the PCA method, four input variables were used in the ANN model. The number of hidden layers and total number of neurons in the hidden layers in an ANN depend on the nature of the problem [38]. Generally, a trial-and-error method is used to obtain a suitable architecture that best reflects the characteristics of the laboratory data. In the present study, an innovative method for calculating the number of neurons in the hidden layers was considered, as shown in the equation below, where NH is the number of neurons in the hidden layers and NI is the number of input variables [39].

$$N_H \leq 2N_I + 1 \tag{10.4}$$

TABLE 10.5
Correlation Matrix for Determining Input Variables by PCA

	Inputs							
Parameter	PCA 1	PCA 2	PCA 3	PCA 4	PCA 5	PCA 6	PCA 7	PCA 8
Eigenvalue	3.2321	2.3445	1.3253	1	0.0677	0.025	0.0055	0
Proportion	0.404	0.293	0.166	0.125	0.008	0.003	0.001	0
Cumulative	0.404	0.697	0.863	0.988	0.996	0.999	1	1

TABLE 10.6
Relationship between Principal Components and Input Variables

Variable	Unit	PCA 1	PCA 2	PCA 3	PCA 4	PCA 5	PCA 6	PCA 7	PCA 8
FA	Mass (%)	−0.377	−0.336	−0.44	0	−0.521	0.113	0.122	0.502
GBFS	Mass (%)	0.349	−0.453	0.303	0	0.046	0.202	−0.628	0.379
CWP	Mass (%)	−0.241	0.307	0.667	0	−0.004	−0.11	0.271	0.563
POFA	Mass (%)	0.361	0.313	−0.503	0	0.46	−0.133	0.045	0.536
SiO_2: Al_2O_3	Ratio	0.33	0.516	−0.007	0	−0.423	0.668	−0.008	0
CaO: SiO_2	Ratio	0.374	−0.473	0.114	0	0.169	0.297	0.712	0
CaO: Al_2O_3	Ratio	0.547	−0.002	0.055	0	−0.554	−0.618	0.092	0
Age	Day	0	0	0	1	0	0	0	0

Since the number of effective input variables is 4, the empirical equation shows that the number of neurons in hidden layers can be less than 9. Therefore, several networks with different topologies, with a maximum of two hidden layers and a maximum of nine neurons, were trained and studied in this study. The hyperbolic tangent stimulation function and Levenberg–Marquardt training algorithm were used in all networks. The statistical indices used to evaluate the performance of different topologies are the root mean squared error (RMSE), average absolute error (AAE), model efficiency (EF) and variance account factor (VAF), which are defined as follows [40]:

$$RMSE = \left[\frac{1}{n}\sum_{i=1}^{n}(P_i - O_i)^2\right]^{\frac{1}{2}} \quad (10.5)$$

$$AAE = \frac{\left|\sum_{i=1}^{n}\frac{(O_i - P_i)}{O_i}\right|}{n} \times 100 \quad (10.6)$$

$$EF = 1 - \frac{\sum_{i=1}^{n}(P_i - O_i)^2}{\sum_{i=1}^{n}(\bar{O}_i - O_i)^2} \quad (10.7)$$

$$VAF = \left[1 - \frac{var(O_i - P_i)}{var(O_i)}\right] \times 100 \quad (10.8)$$

After examining different ANN model topologies, it was found that the network with a 4-5-4-2 topology had the lowest value of error in RMSE, AAE, EF, VAF and the highest value of R2 to estimate the two output parameters. It should be emphasized that the error criteria for training and testing the data are calculated in the main range of variables and not in the normal range. Figure 10.12 illustrates the topology

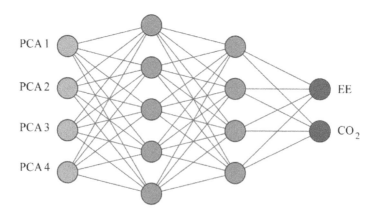

FIGURE 10.12 Topology of a feed-forward ANN with two hidden layers (4–5–4–2 structure).

TABLE 10.7
Properties of COA Parameters

Parameter	Value	Parameter	Value
Number of initial populations	5	Number of clusters that we want to make	1
Minimum number of eggs for each cuckoo	2	Maximum number of cuckoos that can live at the same time	10
Max number of eggs for each cuckoo	10	Control parameter of egg laying	2
Max. iterations of cuckoo algorithm	300		

of a feed-forward ANN network modified by PCA with two hidden layers, four input variables (neurons) and two output parameters.

The ANN used in this study was the Newff Feed Forward. Overall, 70% of the experimental data (118 data), out of 168 experimental data, was used for training, and the remainder 30% (50 data) was used for network testing. To optimize the ANN's weights and biases, the COA was used to provide the least prediction error for the trained structure (modified with PCA). The properties of the COA parameters are shown in Table 10.7. Also, considering that the statistical behaviour of the output data (EE and CO_2) should be evaluated, probability plot diagrams related to determining their normal distribution were examined. The results showed that their statistical behaviour followed a normal distribution, as illustrated in Figure 10.13.

10.6.4 Model Predictions and Results

The results of the trained and optimized PCA-COA-ANN model are depicted in Figures 10.14 and 10.15 for the EE and CO_2 emissions output parameters, respectively. The results indicate that the PCA-COA-ANN estimated reliable and accurate values for the ratio of observational to computational values, R^2, for both input parameters, indicating high accuracy and robustness of the proposed model. Tables 10.8 provides the final weights and biases for both hidden layers estimated by the PCA-COA-ANN model. Using the values of these weights and biases between the different ANN layers, the two output parameters (EE and CO_2 emissions) can be determined and predicted. Moreover, these final weight and bias values can be used to design GPMs with targeted mechanical properties and CO_2 emissions with respect to the availability of industrial by-products and environmental conditions. Accordingly, rather than executing extensive and laborious experimental programmes to reach reasonable results, the trained model could be run in very short time to obtain near optimal results. Only limited experimental validation could be carried out to ensure that variability in local materials and experimental equipment and procedures do not alter the model predictions significantly. Moreover, the experimental validation data could be cumulated and used further in model training and fine tuning for local conditions, which could save time and cost of GPM mixture design development.

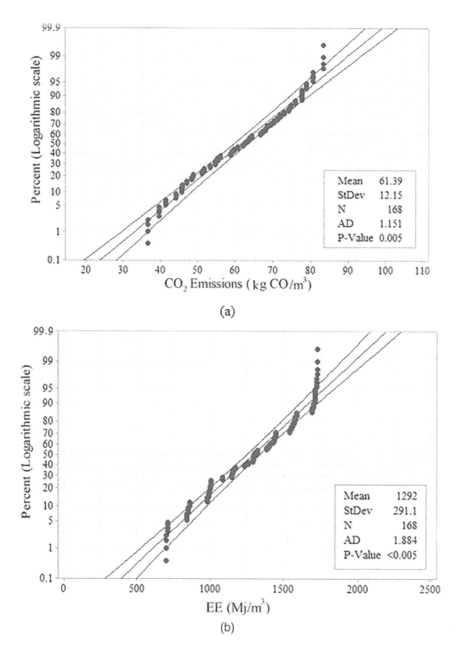

FIGURE 10.13 Probability plot diagrams. (a) CO_2 emissions. (b) EE.

Sustainability of Geopolymer

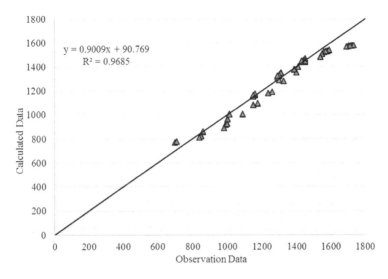

FIGURE 10.14 Predicted versus experimental values of EE estimated by PCA-COA-ANN model.

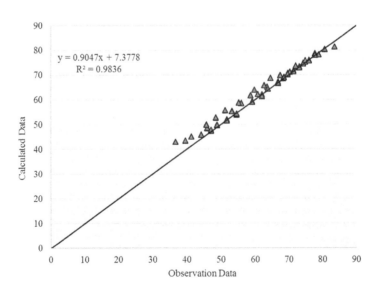

FIGURE 10.15 Predicted versus experimental values of CO_2 emissions estimated by PCA-COA-ANN model.

TABLE 10.8
Final Weight and Bias Values of the Optimum PCA-COA-ANN Model

IW					b1
0.2628	−1.693	0.5162	−1.0867		−2.0935
−1.3895	−1.4424	−0.1549	0.5895		1.0467
−1.045	−1.3043	0.1852	−1.2471		0
−0.9598	0.9747	−1.3313	−0.8596		−1.0467
1.3945	1.006	1.0623	0.5454		2.0935

LW1					b2
1.4828	−0.0699	0.045	0.7636	−0.7899	−1.8473
−1.2436	−0.1904	0.032	−0.3604	1.3034	0.6158
1.0932	−0.1355	0.8086	0.7932	0.9571	0.6158
1.1276	−0.792	1.2058	0.1342	0.2051	1.8473

LW2				b3
0.4349	−1.0378	−0.1033	1.2227	−1.6649
0.3141	−0.7172	−0.9725	−1.1014	1.6649

IW: Weights values for input layer; LW1: Weights values for first hidden; LW2: Weights values for the second hidden layer; b1: Bias values for the first hidden layer; b2: Bias values for the second hidden layer; b3: Bias values for the output layer.

10.7 SUMMARY

This study explored the 'cradle-to-gate' LCA of ternary blended geopolymer mortars composed of industrial by-products with the system's boundary extended to include the mechanical and durability properties of GPM mixture designs on the basis of performance criteria. In addition, using the experimental test database thus developed, an optimized ANN model with PCA was combined with the COA (PCA-COA-ANN) to estimate the CO_2 emission and EE of GPMs. The following main findings can be drawn from this research.

i. The results indicate that the average CS of the studied AAM mixtures was 61.3 MPa, which compares well with traditional cement-based mortars. The highest and lowest mechanical properties were recorded for GPM mixtures made with high content of GBFS and POFA, respectively.

ii. On average, the residual CS and specimen mass declined by 90% and 0.56%, respectively, after 365 days of immersion in the sulphuric acid solution. The GPM mixtures with high GBFS dosage experienced major reduction in CS by an average of 300%. In addition, it was found that GPM mixtures made with high WCP and FA contents provided better resistance to both sulphuric acid and sulphate attack.

iii. UPV exhibited almost a direct relationship with CS for all GPM mixtures tested. However, after immersion in the sulphuric acid and sulphate solutions, the relationship between the pulse velocity and CS followed an irregular pattern, which depended on the dosage of each industrial by-product in the mixture.

iv. The conventional cradle-to-gate LCA revealed that the GPM mixture made with high-volume FA emitted the least amount of CO_2 and consumed the least amount of energy with average values of 45.5 kg CO_2/m^3 and 881.2 MJ/m^3, respectively. However, the AAM mixture made with high-volume GBFS and POFA emitted the highest amount of CO_2 (70.6 kg CO_2/m^3) and consumed the highest amount of energy (1534.5 MJ/m^3), respectively. Nevertheless, these values are significantly lower than that of the benchmark conventional mortar made with pure OPC with 436.8 kg CO_2/m^3 CO_2 emissions and 2793 MJ/m^3 EE.

v. The modified LCA with respect to CS revealed that in GPM mixtures containing high-volume FA and GBFS, lower intensity of normalized CO_2 emissions was achieved by an average of 0.73 $CO_2.m^{-3}$/MPa. However, the highest intensity of normalized CO_2 emissions was achieved by GPM mixtures containing high-volume POFA, with around 1.53 $CO_2.m^{-3}$/MPa, as correlated to its relatively low CS and high amount of electricity required for oven-drying of POFA. The modified LCA that included durability in the performance criteria showed that the normalized CO_2 emission in GPMs containing high-volume POFA and GBFS was relatively higher, with an average intensity of around 3.15 $CO_2.m^{-3}$/MPa, than that of other mixture designs. This issue can be explained by the fact that the mixtures containing GBFS and POFA were vulnerable to sulphuric acid and sulphate attack where their CS decreased significantly after 365 days of immersion in these solutions.

vi. For accurate estimation of the output parameters, considering the total number of input variables, PCA was used to reduce the inputs in the ANN. Moreover, the hyperbolic tangent stimulation function and Levenberg–Marquardt training algorithm were used to determine the best topology for the ANN. Several statistical metrics including RMSE, AAE, EF and VAF were used to evaluate the performance of the proposed ANN topology. The PCA-COA-ANN hybrid model provided satisfactory results to estimate the EE and CO_2 emissions of GPM mixtures, with R^2 values of 0.971 and 0.981 for EE and CO_2 emissions, respectively. Using the optimized weights and biases of the PCA-COA-ANN hybrid model, it is possible to design GPM mixtures with targeted mechanical properties and CO_2 emissions considering the availability of local industrial by-product.

REFERENCES

1. Youn, M.H., et al., Carbon dioxide sequestration process for the cement industry. *Journal of CO_2 Utilization*, 2019. **34**: pp. 325–334.
2. Gupta, N., R. Siddique, and R. Belarbi, Sustainable and greener self-compacting concrete incorporating industrial by-products: a review. *Journal of Cleaner Production*, 2020. **284**: p. 124803.

3. Hwang, C.L., D.H. Vo, and T.P. Huynh, Physical–microstructural evaluation and sulfate resistance of no-cement mortar developed from a ternary binder of industrial by-products. *Environmental Progress & Sustainable Energy*, 2020. **39**(5): p. e13421.
4. Hossain, M.U., et al., Evaluating the environmental impacts of stabilization and solidification technologies for managing hazardous wastes through life cycle assessment: A case study of Hong Kong. *Environment International*, 2020. **145**: p. 106139.
5. Wi, S., et al., Assessment of recycled ceramic-based inorganic insulation for improving energy efficiency and flame retardancy of buildings. *Environment International*, 2019. **130**: p. 104900.
6. Tucker, E.L., et al., Economic and life cycle assessment of recycling municipal glass as a pozzolan in portland cement concrete production. Resources, *Conservation and Recycling*, 2018. **129**: pp. 240–247.
7. Sinka, M., et al., Comparative life cycle assessment of magnesium binders as an alternative for hemp concrete. Resources, *Conservation and Recycling*, 2018. **133**: pp. 288–299.
8. Huseien, G.F. and K.W. Shah, Durability and life cycle evaluation of self-compacting concrete containing fly ash as GBFS replacement with alkali activation. *Construction and Building Materials*, 2020. **235**: p. 117458.
9. Phoo-ngernkham, T., et al., Flexural strength of notched concrete beam filled with alkali-activated binders under different types of alkali solutions. *Construction and Building Materials*, 2016. **127**: pp. 673–678.
10. El Idrissi, A.C., et al., Alkali-activated grouts with incorporated fly ash: from NMR analysis to mechanical properties. *Materials Today Communications*, 2018. **14**: pp. 225–232.
11. Robayo-Salazar, R., et al., Life cycle assessment (LCA) of an alkali-activated binary concrete based on natural volcanic pozzolan: a comparative analysis to OPC concrete. *Construction and Building Materials*, 2018. **176**: pp. 103–111.
12. Bajpai, R., et al., Environmental impact assessment of fly ash and silica fume based geopolymer concrete. *Journal of Cleaner Production*, 2020. **254**: p. 120147.
13. Abdulkareem, M., J. Havukainen, and M. Horttanainen, How environmentally sustainable are fibre reinforced alkali-activated concretes? *Journal of Cleaner Production*, 2019. **236**: p. 117601.
14. Nazer, A., et al., Use of ancient copper slags in Portland cement and alkali activated cement matrices. *Journal of Environmental Management*, 2016. **167**: pp. 115–123.
15. Hou, P., et al., Estimate ecotoxicity characterization factors for chemicals in life cycle assessment using machine learning models. *Environment International*, 2020. **135**: p. 105393.
16. Anastasiou, E., A. Liapis, and I. Papayianni, Comparative life cycle assessment of concrete road pavements using industrial by-products as alternative materials. *Resources, Conservation and Recycling*, 2015. **101**: pp. 1–8.
17. Oyebisi, S., et al., Assessment of activity indexes on the splitting tensile strengthening of geopolymer concrete incorporating supplementary cementitious materials. *Materials Today Communications*, 2020. **24**: p. 101356.
18. Kubba, Z., et al., Effect of sodium silicate content on setting time and mechanical properties of multi blend geopolymer mortars. *Journal of Engineering and Applied Sciences*, 2019. **14**(7): pp. 2262–2267.
19. Dhandapani, Y., et al., Mechanical properties and durability performance of concretes with limestone calcined clay cement (LC3). *Cement and Concrete Research*, 2018. **107**: pp. 136–151.
20. Wang, Y., et al., Utilizing coral waste and metakaolin to produce eco-friendly marine mortar: hydration, mechanical properties and durability. *Journal of Cleaner Production*, 2019. **219**: pp. 763–774.

21. Li, Z., Drying shrinkage prediction of paste containing meta-kaolin and ultrafine fly ash for developing ultra-high performance concrete. *Materials Today Communications*, 2016. **6**: pp. 74–80.
22. Zhang, P., et al., Effects of magnesia expansive agents on the self-healing performance of microcracks in strain-hardening cement-based composites (SHCC). *Materials Today Communications*, 2020. **25**: p. 101421.
23. Standard, A., *C618-15 standard specification for coal fly ash and raw or calcined natural pozzolan for use in concrete*. 2015: ASTM International.
24. Song, X., et al. Durability of fly ash based geopolymer concrete against sulphuric acid attack. in *International conference on durability of building materials and components*. 2005. **10**: 1–12.
25. Trtnik, G., F. Kavčič, and G. Turk, Prediction of concrete strength using ultrasonic pulse velocity and artificial neural networks. *Ultrasonics*, 2009. **49**(1): pp. 53–60.
26. Jeon, I.K., et al., Investigation of sulfuric acid attack upon cement mortars containing silicon carbide powder. *Powder Technology*, 2020. **359**: pp. 181–189.
27. García-Vera, V.E., et al., Influence of crystalline admixtures on the short-term behaviour of mortars exposed to sulphuric acid. *Materials*, 2019. **12**(1): p. 82.
28. Hammond, G., et al., *Inventory of carbon & energy: ICE*. Vol. 5. 2008: Sustainable Energy Research Team, Department of Mechanical Engineering.
29. Teixeira, E., et al., Quality and durability properties and life-cycle assessment of high volume biomass fly ash mortar. *Construction and Building Materials*, 2019. **197**: pp. 195–207.
30. Li, J., et al., Eco-friendly mortar with high-volume diatomite and fly ash: Performance and life-cycle assessment with regional variability. *Journal of Cleaner Production*, 2020: p. 121224.
31. AzariJafari, H., et al., Ternary blended cement: an eco-friendly alternative to improve resistivity of high-performance self-consolidating concrete against elevated temperature. *Journal of Cleaner Production*, 2019. **223**: pp. 575–586.
32. Gursel, A.P., H. Maryman, and C. Ostertag, A life-cycle approach to environmental, mechanical, and durability properties of "green" concrete mixes with rice husk ash. *Journal of Cleaner Production*, 2016. **112**: pp. 823–836.
33. Nikoo, M., et al., Determining the natural frequency of cantilever beams using ANN and heuristic search. *Applied Artificial Intelligence*, 2018. **32**(3): pp. 309–334.
34. Haykin, S., *Neural networks: a comprehensive foundation*. 2007: Prentice-Hall, Inc.
35. Bishop, C.M., *Pattern recognition and machine learning*. 2006: springer.
36. Rajabioun, R., Cuckoo optimization algorithm. *Applied Soft Computing*, 2011. **11**(8): pp. 5508–5518.
37. Jolliffe, I.T., Graphical representation of data using principal components. *Principal Component Analysis*, 2002. **1**: pp. 78–110.
38. Hubbard, R. and S.J. Allen, An empirical comparison of alternative methods for principal component extraction. *Journal of Business Research*, 1987. **15**(2): pp. 173–190.
39. Faridmehr, I., M. L. Nehdi, M. Nikoo, G. F. Huseien, and T. Ozbakkaloglu, Life-cycle assessment of alkali-activated materials incorporating industrial byproducts. *Materials*, 2021. **14**(9): p. 2401.
40. Alabduljabbar, H., G. F. Huseien, A. R. M. Sam, R. Alyouef, H. A. Algaifi, and A. Alaskar, Engineering properties of waste sawdust-based lightweight alkali-activated concrete: experimental assessment and numerical prediction. *Materials*, 2020. **13**(23): p. 5490.

Index

abrasion 2, 5, 19, 21, 52, 67, 124
aggressive environment 1, 5, 20, 33, 97, 123
alkali-activated mortars 32, 98, 101, 157
alkaline solution 7, 12, 17, 23, 31, 39
aluminium 6, 15, 42, 57, 62, 79, 84, 99

bending stress 18, 64, 71, 99, 116, 118
binder 1, 6, 12, 14, 19, 23, 35, 42, 124
bond strength 5, 12, 16, 52, 59, 65, 78

calcium 2, 7, 12, 17, 31, 42, 51, 156, 174
cement 1, 3, 5, 7, 16, 19, 23, 42, 56
cementitious 71, 80, 99, 149, 170, 179, 190
climate change 1, 4, 189
commercial 3, 5, 11, 20, 23, 61, 77, 97, 167
compatibility 5, 97, 99, 99, 103, 113, 170
compressive strength 5, 18, 31, 54, 87, 174
concrete 1–6, 16, 33, 61, 78, 99, 124
construction 1, 7, 21, 31, 97, 123, 150
cracks 4, 17, 21, 34, 71, 104, 136, 168, 187
crystallization 4, 167

damage 1, 3, 97, 118, 152, 158, 163, 176
deformation 4, 104
degradation 2, 4, 16, 23, 123, 167, 169, 186
deterioration 1, 3, 20, 37, 98, 130, 157, 187
development 5, 7, 12, 34, 40, 57, 97
durability 1, 6, 17, 23, 33, 98, 124, 150, 199

eco-friendly 27, 175, 204
economic 1, 21, 167
economically 7, 23
economic benefits 24, 124, 180
efficiency 16, 125, 167, 175, 198
environment 1–7, 15, 21, 37, 63, 123, 163

fly ash 7, 77, 97, 119, 149, 177

gels 7, 13, 32, 55, 69, 88, 138
geopolymer 6, 42, 151, 156, 175, 189
geopolymerization 11, 14, 39, 56, 67, 117
green concrete 7, 19, 59, 77, 179, 189
greenhouse 1, 6, 22, 149, 175

harsh environment 3, 167
high performance 6, 23, 42, 61, 97, 102, 182
hydration 3, 5, 17, 20, 38, 59, 83, 106

immersion 134–142, 184–188, 193, 202, 203
improvement 2, 13, 18, 23, 98, 106, 161, 176
infrastructure 2, 167
initial 35, 53, 84

landfill 37, 42, 123, 149
life cycle assessment 180, 187

maintenance 1–4, 31, 97, 124, 167, 170
microstructures 14, 68, 118, 130, 136, 160
mineral 6, 7, 77, 170
molarity 6, 17, 42, 54, 61, 72, 85, 101, 151
mortar 1–7, 16, 23, 34, 37, 54, 78, 104

nanomaterials 3, 178
natural 1, 12, 22, 37, 42, 99, 123, 149, 179

permeability 5, 16, 38, 171, 175
polymer 1, 5, 11, 13, 24
porosity 2, 17, 36, 106, 124, 152, 163, 172
potential 7, 11, 15, 24, 78, 175, 189
problem 1, 6, 31, 97, 123, 149, 168, 176, 197

reaction 2, 4, 13, 18, 35, 59, 70, 104
repair 1, 3, 5, 11, 19, 56, 68, 72, 90
resistance 1, 5, 24, 33, 67, 97, 130

setting time 13, 24, 36, 53, 77, 86, 102, 118
shrinkage 5, 11, 38, 51, 65, 102, 160, 170
silica 6, 12, 20, 33, 54, 61, 73, 88, 126, 132
slag 7, 15, 24, 36, 53, 77, 103, 124, 150
sodium hydroxide 6, 19, 23, 52, 58, 101, 189
sodium silicate 6, 21, 54, 60, 101, 128, 182
strength 5, 12, 17, 87, 102, 110, 163, 175
sulphate 2, 4, 17, 21, 37, 124, 137, 141, 203
sulphuric 20, 33, 97, 125, 134, 184, 194, 202
surfaces 1, 5, 68, 73, 102, 158, 170
sustainability 6, 23, 56, 124, 128, 179

thawing-freezing 2, 4, 158, 164

urbanization 2, 4

wastes 7, 12, 16, 31, 37, 42, 97, 123, 149
workability 16, 23, 32, 39, 62, 81, 103, 180